Eric Emmet

?

The Penguin Book of
Brainteasers

Compiled by David Hall
and Alan Summers

Viking

VIKING
Penguin Books Ltd, Harmondsworth, Middlesex, England
Viking Penguin Inc., 40 West 23rd Street, New York, New York 10010, U.S.A.
Penguin Books Australia Ltd, Ringwood, Victoria, Australia
Penguin Books Canada Ltd, 2801 John Street, Markham, Ontario, Canada L3R 1B4
Penguin Books (N.Z.) Ltd, 182–190 Wairau Road, Auckland 10, New Zealand

First published 1984

Reproduced, printed and bound in Great Britain by
Hazell Watson & Viney Limited,
Member of the BPCC Group,
Aylesbury, Bucks
Set in 9/11pt Linotron Trump by
Rowland Phototypesetting Ltd,
Bury St Edmunds, Suffolk

British Library Catalogue in Publication Data

Emmet, E. R.
    The Penguin book of brainteasers.
    1. Mathematical recreations
    I. Title   II. Hall, David
    III. Summers, Alan
    793.7'4   QA95

    ISBN 0-670-80066-X

Contents

Preface

Readers of Eric Emmet's previous books will have been sad to hear of his death in March 1980, but will be glad to learn that he left a wealth of unpublished puzzles behind him. Eric thoroughly enjoyed making up puzzles – this is reflected in the number he produced – but not just for the sake of it: he wanted to help people on the path to clear thinking. The thorough answers are a good example of the logical route to the solution and the infinite care Eric took to achieve the unique result.

We have compiled this book from the many original puzzles and, in doing so, have attempted to maintain a format of which, we hope, Eric would have approved. Between us, over a number of years, we have checked a great many puzzles for Eric and we are very glad to be able to continue with this book as Eric had planned.

There are cross numbers, addition, multiplication and division sums with letters substituted for numbers, some complicated by missing letters and others further complicated by a few, or all, letters being wrong. Eric particularly enjoyed contriving football and cricket puzzles and there are a number of these, too.

Of the many types of puzzle of varying difficulty, all require the clear thinking approach which Eric fostered. It may help the new-comer in tackling these puzzles to have a look at a solution and follow the technique used – the key to many a problem is in finding out where to begin. The solutions are, however, not the only path to the correct answer but merely a guide.

Many of Eric's pupils will remember trying out his latest brain-teaser in Friday afternoon lessons. No better encouragement was given than to offer a financial reward (related, of course, to difficulty) for the first correct solution through his letter box before ten on Saturday morning. Friday evenings were uncommonly busy.

After compilation of any book, and especially one of this nature, comes the task of checking, for which we are particularly grateful to Nick Taylor-Young who has been through all the puzzles. Again, for a book of this kind, accurate typing is essential and has been admirably undertaken by Eleanor Rose and Sarah Mais. In addition, much help in checking and preparation of the copy has been given by Alex Summers.

Aileen Emmet had always been a tremendous support to Eric and her loss left a great void in the last few months of Eric's life. It is to the indefatigable team of Aileen and Eric, as we remember them, that we dedicate this book.

Part I

?

Cross Number
Puzzles

1. Greater or Lesser

<table>
<tr><td>1</td><td></td><td>2</td></tr>
<tr><td>3</td><td>4</td><td></td></tr>
<tr><td>5</td><td></td><td></td></tr>
</table>

(There are no zeros.)

Across 1. 3 times 3 across.
      3. A multiple of 4 down.
      5. Either each digit is greater than the one before or each digit is less than the one before.

Down 1. Sum of digits is 18.
      2. A multiple of 3.
      4. The square of an odd number.

2. A Cross Number (3 by 3)

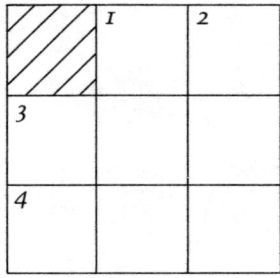

(There are no zeros.)

Across 1. The sum of the digits is 10.
 3. Digits all even.
 4. Digits all odd, and each one is less than the one before.

Down 1. The second digit is greater than either of the other two.
 2. A multiple of 3 down.
 3. The second digit is greater than the first one.

(One of these numbers is the same as another one reversed.)

3. A Cross Number (5 by 5)

1	2	3	4	
5		6	▨	
7		▨	8	9
▨	10	11	▨	
▨	12			

(There are no zeros.)

Across
1. All digits are different, and all odd.
5. The sum of the digits is 10.
6. A perfect square.
7. The sum of the digits is the same as the sum of the first two digits of 9 down.
8. A multiple of 13.
10. A multiple of 6.
12. An even number, which is the same when reversed.

Down
1. 8 across multiplied by 10 or less.
2. Digits all different and in descending order.
3. 8 across reversed.
4. Each digit is greater than the one before.
9. Each digit is less than the one before.
11. The same when reversed.

4. A Cross Clue

1	2	3
4		
▨	5	

(There are no zeros.)

Across 1. Each digit is odd and is greater than the one before.
      4. The digits are all different and this is a multiple of the number which is 3 greater than 1 down. Even when reversed.
      5. A perfect cube.

Down 1. 17 goes into this.
     2. A multiple of 1 down.
     3. Each digit is odd and is less than the one before.

One clue is incorrect, which one?
With which digit should each square be filled?

5. Some Very Cross Numbers

It might be said – indeed it has been said – that Uncle Bungle has a genius for getting things wrong. This has not hitherto ruffled the imperturbability of his temper, but the other day, perhaps as a result of excessive pulling of a leg that has perceptibly lengthened through the years, something snapped.

'All right,' he said crossly, 'if you say I get things wrong I'll get things wrong.' And he proceeded to produce a cross number puzzle in which every single clue was incorrect. And this time there was no mistake in the mistakes.

The puzzle was as follows:

1	2	3
4	5	6
7	8	9

(There are no zeros.)

Across 1–2–3. A multiple of neither 3 nor 5 nor 7.
      4–5–6. The sum of digits is greater than 5.
      7–8–9. The last digit is less than the sum of the first two.

Down 1–4–7. The sum of the digits is less than 19.
      2–5–8. The sum of the digits is greater than the sum of the digits of 3 down.
      3–6. An odd number.

Find Uncle Bungle's solution.

6. Odds and Evens

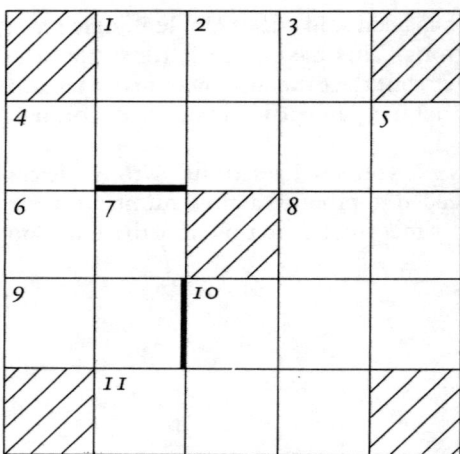

(There are no zeros.)

Across
   1. A multiple of 6 across.
   4. Digits all odd and all different.
   6. Even.
   8. A multiple of the cube root of 4 down.
   9. A prime number.
  10. Digits all odd and all different.
  11. The same when reversed.

Down
   1. A perfect square.
   2. A factor or a multiple of 10 down.
   3. Each digit is greater than the preceding one.
   4. A perfect cube.
   5. The digits are in descending order.
   7. A perfect square.
  10. A multiple or a factor of 2 down.

7. A Crosser Number

(There are no zeros.)
One of the clues which follow is incorrect. Which one?

Across   1. Each digit is less than the one before.
          4. A prime factor of the last 2 digits of 8 across.
          5. A multiple of 9.
          6. Digits all even, and all different.
          8. The sum of the digits is the same as the sum of the digits of 2 down.
          9. A multiple of 19.
       10. Twice the square of the cube root of 3 down.

Down   1. The last digit is equal to the difference between the sum of the first and second digits and the sum of the third and fourth digits.
          2. The same when reversed.
          3. A perfect cube.

7. A perfect square when reversed.
8. Even.

Find out as much as you can of the correct solution.

Part II

?

Cricket

8. One Good Score

In a cricket competition four teams – A, B, C and D – all play each other once. Points are awarded as follows:

To the side that wins: 10.

To the side that wins on the first innings in a drawn match: 6.

To the side that loses on the first innings in a drawn match: 2.

To each side for a tie: 5.

To the side that loses: 0.

A, B, C and D get 16, 11, 21 and 6 points respectively, and you are told that A won 1 match.

Find the result of each match.

9. One Winner

A, B, C and D have all played each other once at cricket. Points are awarded as follows:

To the side that wins: 10.

To the side that wins on the first innings in a drawn match: 6.

To the side that loses on the first innings in a drawn match: 2.

To each side for a tie: 5.

To the side that loses: 0.

A, B, C and D got 10, 11, 17 and 14 points respectively, and you are told that only 1 match was won outright.

Find the result of each match.

10. Cricket: Four Teams

A, B, C and D are all to play each other once at cricket. After some – or possibly all – of the matches have been played, A had got 18 points, B had got 17 and C had got 21. I'm afraid, however, that I was not able to find out how many points D had got. Points are awarded as follows:

To the side that wins: 10.

To the side that wins on the first innings in a drawn match: 6.

To the side that loses on the first innings in a drawn match: 2.

To each side for a tie: 5.

To the side that loses: 0.

Find the results of all the matches that were played.

11. Latest Results

A, B, C and D are having a cricket competition with each other in which eventually they are all going to play each other once. Points are awarded as follows:

To the side that wins: 10.

To the side that wins on the first innings in a drawn match: 6.

To the side that loses on the first innings in a drawn match: 2.

To each side for a tie: 5.

To the side that loses: o.

The latest news I have about their points is as follows:

A: 21

B: 10

C: 9

D: 6.

Find the result of each match.

12. And Where Was B?

A, B, C and D have all played each other once at cricket. Points are awarded as follows:

To the side that wins: 10.

To the side that wins on the first innings in a drawn match: 6.

To the side that loses on the first innings in a drawn match: 2.

To each side for a tie: 5.

To the side that loses: 0.

I am able to reveal the fact that A, C and D got 18, 17 and 10 points respectively, but all that I can tell you about B's points is that they got more than 7 and less than 11.

Find the result of each match.

Part III

?

Football

13. Can You Draw Any Conclusions?

Four football teams – A, B, C and D – are to play each other once. After some – or perhaps all – of the matches have been played a document giving some details of matches played, won, lost, drawn, goals for, goals against and points looked like this:

	PLAYED	WON	LOST	DRAWN	GOALS FOR	GOALS AGAINST	POINTS
A					4	4	
B					5		5
C		0				4	2
D					0	3	0

(2 points are given for a win, and 1 point to each side for a draw.)

Find the score in each match.

14. C's Loss

Five football teams – A, B, C, D and E – are each to play each other once.

After some of the matches have been played a table giving some details of matches played, won, lost, etc., looks like this:

	PLAYED	WON	LOST	DRAWN	GOALS FOR	GOALS AGAINST
A				0	1	1
B	4		1		9	5
C	1				1	3
D			0	0	2	1
E						

Can you find the score in each match?

15. Four Teams: Letters for Digits, One Letter Wrong

Four football teams – A, B, C and D – are to play each other once. After some – or perhaps all – of the matches had been played a table was drawn up giving some details of the matches played, won, lost, etc. But when I say a table perhaps that was rather misleading, for not only had the digits (from 0 to 9) been replaced by letters, but also one of the letters was incorrect on one of the occasions on which it appeared (if it appeared more than once). Each letter should stand for the same digit wherever it appeared and different letters should stand for different digits.

The table looked like this:

	PLAYED	WON	LOST	DRAWN	GOALS FOR	GOALS AGAINST	POINTS
A				t	h	h	t
B	x			p	y	y	x
C	y	x			g	h	x
D			x		p	h	p

(2 points are given for a win and 1 point to each side in a drawn match.)

Which letter was wrong? What should it be? Find the score in each match.

16. Four Teams

Four football teams are to play each other once. After some of the matches have been played a document giving some details of the matches played, won, lost, etc., looked like this:

	PLAYED	WON	LOST	DRAWN	GOALS FOR	GOALS AGAINST	POINTS
A	2					3	
B						4	0
C				1	7	5	
D		2			3	3	

(2 points are given for a win and 1 point to each side in a drawn match.)

Find the score in each match.

17. You Can't Do One Without the Other

In the following football table and addition sum, letters have been substituted for digits (from 0 to 9). The same letter stands for the same digit wherever it appears and different letters stand for different digits.

The three teams are eventually going to play each other once – or perhaps they have already done so.

(i)

	PLAYED	WON	LOST	DRAWN	GOALS FOR	GOALS AGAINST	POINTS
A					h	m	d
B					m	y	t
C		p			p	t	x

(ii)
$$\begin{array}{r} t\ \ d \\ d\ \ x \\ \hline m\ \ t \end{array}$$

(2 points are given for a win and 1 point to each side in a drawn match.)

Find the scores in the football matches and write out the addition sum with numbers substituted for letters.

18. Return Matches

Three teams, A, B and C, are each to play each other at football twice. After some of the matches had been played it was interesting to notice that if two teams had completed their 2 matches against each other, neither of them scored the same number of goals in the second match as they did in the first. It is also interesting to record that in no 2 matches were the scores exactly the same.

I have set out below all that I was able to discover about the details.

	PLAYED	WON	LOST	DRAWN	GOALS FOR	GOALS AGAINST
A		1		1	3	3
B				0	0	4
C	3					

Find the score in each match.

19. D's Gain

Five football teams – A, B, C, D and E – are all going to play each other once.

After some of the matches have been played a table giving some details of matches played, won, lost, etc., looked like this:

	PLAYED	WON	LOST	DRAWN	GOALS FOR	GOALS AGAINST
A		0		0	3	4
B						
C	3			1	4	2
D	1				2	1
E				1	3	2

Find the score in each match.

20. Football and Addition: Letters for Digits

In the following football table and addition sum, letters have been substituted for digits (from 0 to 9). The same letter stands for the same digit wherever it appears and different letters stand for different digits.

The four teams are eventually going to play each other once.

(i)

	PLAYED	WON	LOST	DRAWN	GOALS FOR	GOALS AGAINST	POINTS
A	m					h	t
B					s	m	g
C		y				x	
D	g			y			m

(ii)

$$
\begin{array}{rr}
x & s \\
x & s \\
\hline
p & g \\
\end{array}
$$

(2 points are given for a win and 1 point to each side in a drawn match.)

Find the scores in the football matches and write out the addition sum with numbers substituted for letters.

21. Seven Teams

Seven football teams – A, B, C, D, E, F and G – are to play each other once. After some of the matches have been played a table giving some details of the matches played, won, lost, etc., looked like this:

	PLAYED	WON	LOST	DRAWN	GOALS FOR	GOALS AGAINST	POINTS
A	1				4	2	
B					3		0
C			0		5	5	3
D			2			8	3
E	2				6	3	4
F	2				0	8	
G	1					3	

(2 points are given for a win and 1 point to each side in a drawn match.)

Find the score in each match.

Part IV

?

Football:
New Method

22. Everyone Scores

The new football competition, in which goals are rewarded as well as wins or draws, has been proving a great success. In the latest competition between four teams, each side scored at least 1 goal every time they played, though not more than 5 goals were scored in any match.

In this competition 10 points are awarded for a win, 5 points for a draw, and 1 point for each goal scored. After all the matches except 1 had been played the situation for the four teams was as follows:

A: 26 points
B: 4 points
C: 18 points
D: 20 points.

It was interesting to notice that D had lost none, and that every team had a different number of goals scored against them.

What was the score in each match?

23. Three Teams

Three teams, A, B and C, have each played each other once at football. 10 points are given for a win, 5 points for a draw, and 1 point for each goal scored. It is hoped that this new method will result in more goals being scored and therefore more attractive football and bigger crowds. It seems that this may be happening to some extent, for in this competition each side scored at least 1 goal in every match.

Their points were:

A: 25
B: 8
C: 9.

Find the score in each match.

24. Three More Teams

Three teams, A, B and C, are all to play each other once at football. 10 points are given for a win, 5 points for a draw and 1 point for each goal scored, whatever the result of the match. After some, or perhaps all, of the matches have been played the points were as follows:

A: 21
B: 20
C: 4.

Not more than 6 goals were scored in any match.

What was the score in each match?

25. All Scores Different

The great thing about the new football method is that goals count whoever wins the match. As a result – for goals are what the game is about – there has been a surge of enthusiasm and a considerable increase in the number of goals.

In this new method 10 points are given for a win, 5 points to each side for a draw and 1 point for each goal scored. In a recent competition between four teams – A, B, C and D – who played each other once, the points were as follows:

A: 25
B: 34
C: 6
D: 20.

Each side scored at least 1 goal in each match, and – rather interestingly – the score was different in every match.

Find the score in each match.

26. The Professor's Versatility

I have known for a long time – and so have his other admirers – about Professor Knowall's interest in soccer. But I must admit that I had not expected him to know about the new soccer experiments designed to produce more goals and attract bigger crowds.

In this new method, 10 points are awarded for a win, 5 points for a draw, and 1 point for each goal scored.

I found the great man looking at the points obtained by four teams who had all played each other once. They read as follows:

A: 8
B: 19
C: 8
D: 57.

'Why is everybody else so careless,' he said, 'including *you*, my dear Sergeant Simple?'

'One of these figures is wrong. But,' he went on, 'if I give you the information that it is possible to discover which one, then it will be possible for you to discover it.'

This cryptic remark was a bit too much for me. But, in the hope that it may not be too much for some of my readers, I pass the information on. Two things that I do know are that every side scored at least 1 goal in each game, and that not more than 10 goals were scored in any match.

Which figure was wrong? And what information can you give about the scores in each match?

27. More Goals

The new method of rewarding goals scored in football matches goes from strength to strength. In this method 10 points are given for a win, 5 points for a draw and 1 point for each goal scored. One can get some idea of the success of the method from the fact that in the latest competition between 5 teams, when some of the matches had been played, each team had scored at least 1 goal in every match. They are eventually going to play each other once.

The points were as follows:

A: 11
B: 8
C: 12
D: 5
E: 43.

Not more than 9 goals were scored in any match.

What was the score in each match?

28. Goals Rewarded Rewarded

The new method of rewarding goals in soccer matches has now been improved still further. Under this method 10 points are given for a win, 5 points for a draw, and 1 point for each goal scored. And now there is to be a bonus point for 2 successive goals, 3 bonus points for 3 successive goals and 5 bonus points for 4 successive goals, and so on.

Three teams, A, B and C, have each played each other once. Their total points, including bonuses, are A: 16; B: 20; and C: 11. Each side got a bonus and there was 1 bonus (but only 1) in every match.

Both sides scored at least 1 goal in every match.

What were the scores in each match?

Part V

?

Figures All Wrong:
Division and Addition

29. Four Digits Divided by Two Digits

In the following, obviously incorrect, division sum the pattern is correct, but all the figures are wrong.

```
        23
  38 ) 2192
      210
       98
       94
```

The correct division comes out exactly. The digits in the *answer* are only 1 out, but all the other digits may be incorrect by any amount.

Find the correct figures.

30. Five Digits Divided by Two Digits

In the following, obviously incorrect, division sum the pattern is correct, but every single figure is wrong.

```
        342
   39 ) 22498
        215
         99
         92
         86
         75
```

The correct division comes out exactly. The digits in the *answer* are only 1 out, but all the other digits may be incorrect by any amount.

Find the correct figures.

31. Addition: Four Numbers

In the following addition sum all the digits are wrong. But the same wrong digit stands for the same correct digit wherever it appears, and the same correct digit is always represented by the same wrong digit.

```
462
264
152
201
───
837
```

Find the correct addition sum.

32. Five Digits Divided by Two Digits

In the following, obviously incorrect, division sum the pattern is correct, but every single figure is wrong.

```
        271
   13 ) 20634
        80
       263
       258
        54
        46
```

The correct division, of course, comes out exactly. All the digits in the *answer* are only 1 out, but all the other digits may be incorrect by any amount.

Find the correct figures.

33. Addition: Two Numbers

Each digit in the addition sum below is wrong. But the same wrong digit stands for the same correct digit wherever it appears, and the same correct digit is always represented by the same wrong digit.

```
 4751
 9731
46082
```

Find the correct addition sum.

34. Only One Out

In the following, obviously incorrect, division sum the pattern is correct, but all the figures are 1 out, that is, 1 more or 1 less than the correct figures. The sum comes out exactly.

```
        71
45 ) 3343
     125
      83
      81
```

Find the correct figures.

Part VI

?

Uncle Bungle's Puzzles

35. Nearly Right

'Well, it's *nearly* right,' said Uncle Bungle as he produced one of his tattered pieces of paper purporting to contain some details of the football matches that had been played between five teams who were all to play each other once.

The figures that could be read were as follows:

	PLAYED	WON	LOST	DRAWN	GOALS FOR	GOALS AGAINST	POINTS
A	3		1	2			2
B	4	3		1	5		7
C		1	1			2	
D	2		2		3	5	0
E	3		2	0	3	2	2

(2 points are given for a win, and 1 point to each side for a draw.)

In fact I discovered subsequently that only *one* of the figures was wrong. *Which one?*

Find the scores in all the matches that had been played.

36. One Wrong Again

Some of the figures in Uncle Bungle's latest puzzle were illegible and some were just not there. I did the best I could to make everything right for his public, but I'm sorry to say that I failed again. Four football teams were to play each other once, and the figures that I was given about the number of matches played, won, lost, etc., looked like this:

	PLAYED	WON	LOST	DRAWN	GOALS FOR	GOALS AGAINST	POINTS
A	3					3	5
B			0	2	5	3	4
C	2	0			6	8	
D	3					8	0

(2 points are given for a win and 1 point to each side for a draw.)

It was only later that I discovered that one of the figures was wrong, so that we do not in fact know whether all the matches had been played or not.

Find the score in each match.

37. Football: Five Teams, Two Figures Wrong

Five football teams are to play each other once. After some of the matches had been played a document giving some details of the matches played, won, lost, etc., was found. But unfortunately Uncle Bungle had been messing about with it, and *two* of the figures were wrong.

Here is the document that was found:

	PLAYED	WON	LOST	DRAWN	GOALS FOR	GOALS AGAINST	POINTS
A	3		0	2	4	4	
B			3		3	5	
C	2			0	5	2	2
D		2		1		5	6
E	4		2	1		5	

(2 points are given for a win and 1 point to each side in a drawn match.)

Find the score in each match.

38. Uncle Bungle and the Vertical Tear

It was, I'm afraid, typical of Uncle Bungle that he should have torn up the sheet of paper which gave particulars of the number of matches played, won, lost, drawn, etc., of three local football teams who were eventually going to play each other once. Not only had Uncle Bungle torn it up, but he had also thrown half of it on to the fire which seems to burn eternally in his grate. The tear was a vertical one and the only things that were left were the goals for, the goals against, and the points.

The situation was not made any easier by the fact that the digits had been replaced by letters.

What was left was as follows:

GOALS FOR	GOALS AGAINST	POINTS
m	p	m
x	x	p
g	x	x

Calling the teams A, B and C, in that order, *find the score in each match.*

39. Uncle Bungle and the Horizontal Tear

Last time it was vertical, but no one could accuse Uncle Bungle of being consistent and this time it was horizontal. I mean the way in which he tore the piece of paper on which were written the details of the matches which four local football teams, A, B, C and D, had been playing against each other.

All that was left was:

	PLAYED	WON	LOST	DRAWN	GOALS FOR	GOALS AGAINST
A	3	1	0	2	7	5
B	3	2	1	0	5	5

It is known that there was 1 match still to be played. Not more than 7 goals were scored in any game.

With the information that it is possible to discover the score in each match you should be able to discover it.

What was the score in each match?

40. The Lie Drug

As my readers will know, Uncle Bungle is always in the forefront of any new idea. He has been experimenting recently with a Lie Drug, which is designed to make people's remarks untrue, and he has combined this with his lifelong interest in football to produce some rather interesting results.

The secretaries of four local football teams, whom we shall call A, B, C and D, were given a dose of this drug and then asked to write down the number of matches played, won, lost, drawn, the goals for, the goals against and the points of their teams, who were at the time engaged in a competition with each other. It was interesting to notice that the drug was completely successful in making all their figures false – they were, in fact, all 1 out, that is, 1 more or 1 less than the correct number. But whether this was by accident or by design on the part of the maker of the drug, I would not know.

The figures that the four secretaries produced were as follows:

	PLAYED	WON	LOST	DRAWN	GOALS FOR	GOALS AGAINST	POINTS
A	2	0	0	0	6	7	4
B	1	0	1	0	0	1	4
C	3	1	0	0	3	3	2
D	2	1	1	0	4	2	0

(2 points are given for a win and 1 point for a draw.)

Find the score in all the matches played.

41. Uncle Bungle and a Lucky 13

Uncle Bungle is always, as my readers will know, on the lookout for new ideas. The puzzle that he has just made up is his first long multiplication sum ever. But (what happened to him?) there are no mistakes. As it is the first sum of its kind, my uncle thought it would be a good idea to give the information that the final result is divisible by 13.

The letters stand for digits (from 0 to 9) and the same letter stands for the same digit wherever it appears.

$$
\begin{array}{rrrrrr}
  &   & - & - & - & - \\
  &   &   &   & - & t \\
\hline
  & p & x & k & y & p \\
  & r & y & d & x &   \\
\hline
p & p & t & t & m & p \\
\end{array}
$$

Write out the complete sum.

42. No Divisor

Uncle Bungle has been doing a division sum with letters substituted for digits. But unfortunately, whether by accident or design, he has left out the divisor.

What was left looked like this:

```
          g  x  a  p  m  g
       ┌─────────────────────
       )  m  m  d  p  b  p  g
          m  g
          ─────
             g  d  p
             g  m  a
             ────────
                b  b
                b  p
                ─────
                   p  p
                   p  m
                   ─────
                      m  g
                      m  g
                      ─────
```

Find the divisor and all the digits of the sum.

43. Addition: Letters for Digits, One Letter Wrong

Uncle Bungle has recently become very keen on making addition sums in which the digits are replaced by letters. In these sums the same letter should stand for the same digit wherever it appears, and different letters should stand for different digits.

Notice that I say 'should'. Those who know him well will not be surprised to hear that my uncle has bungled once more, and that *one* of the letters in the sum below is wrong.

His sum looked like this:

```
T L S L T T E M A
T L S T B T T M K
B B S H L M D S A
```

What can you say about the letter that was wrong?

Find the correct sum.

44. Uncle Bungle Gets the Last Line Wrong

'If you know what you are adding you should be able to add it,' said Uncle Bungle when I complained to him that though the first two lines of his addition sum were legible, the last line across was not. I can, of course, see what he means by this, but he seems to forget that in his addition sum letters have been substituted for digits. The same letter stands for the same digit whenever it appears and different letters stand for different digits.

In the last line across, however, Uncle went haywire, and since four of the letters were illegible I have replaced them by blanks. The sum reads like this:

```
  E X M R E E K
  E H K R E K K
- K - H - X - E
```

(From the remark that Uncle made I gathered that in the final sum all 10 digits (0 to 9) appear.)

Find the correct addition sum.

45. Bungled Again

In this addition sum, with letters substituted for digits, there is, I'm afraid, a mistake. The first two lines across are correct, but in the third line across there has been a mistake and one of the letters is incorrect. Apart from this unfortunate error (due, of course, to Uncle Bungle), each letter stands for the same digit wherever it appears and different letters stand for different digits.

```
   H  B  B  D  B  M  D  B
   H  B  B  B  P  A  D  B
 M B  H E  G  T  X  B  H
```

Find the mistake and write out the correct addition sum.

Part VII

?

Letters for Digits:
Division, Addition and
Multiplication

46. H is 3

In the following multiplication each letter stands for a different digit. You are told that H is 3.

```
  Y  X  P
        H
---------
P  M  Y  X
```

Write out the sum with numbers substituted for letters.

47. Rows of Eight

In the following addition sum the digits have been replaced by letters. The same letter stands for the same digit wherever it appears and different letters stand for different digits.

```
D H G D N G N C
C H C D N D G D
E P B N B B G E
```

Find the digits for which the letters stand.

48. Rows of Nine

In the addition sum below the digits have been replaced by letters. The same letter stands for the same digit wherever it appears, and different letters stand for different digits.

```
M M W X T F G G G
M M E X W T F G G
M M Y F M M F G G
F T T Y M C V F M
```

Find the digits for which the letters stand.

49. Multiplication of Seven Digits

In the multiplication sum below digits have been replaced by letters.
The same letter stands for the same digit wherever it appears, and
different letters stand for different digits.

```
  A T V T S A V
            A
---------------
V V J E C K S A
```

You are told that A is not greater than 5.

Find the digits for which the letters stand.

50. Two Rows

In the following addition sum the digits have been replaced by letters. The same letter stands for the same digit wherever it appears, and different letters stand for different digits.

```
  A X S X V A E
  X S S A V A E
  W H D T X E E
```

Find the digits for which the letters stand.

51. Five Digits Divided by Two Digits

In the following division sum each letter stands for a different digit.

```
              m  a  h  q
        a  g ) d  b  a  r  p
              m  q
              ─────
                 g  a
                 a  g
                 ─────
                    j  r
                    q  m
                    ─────
                       j  p
                       j  p
                       ─────
```

Find the division sum with all the letters replaced by digits.

52. Addition: Two Numbers

Below is an addition sum with letters substituted for digits. The same letter stands for the same digit whenever it appears, and different letters stand for different digits.

```
   X D
 H H D
 X D H
```

Write the sum out with numbers substituted for letters.

53. Multiplication of Five Digits

In the multiplication sum below the digits have been replaced by letters. The same letter stands for the same digit wherever it appears, and different letters stand for different digits.

```
R E M H B
        M
─────────
T M G B R
```

Write the sum out with letters replaced by digits.

54. More Addition-by-Letters

In the addition sum below the digits have been replaced by letters.
The same letter stands for the same digit wherever it appears, and
different letters stand for different digits.

```
A A R X B A A
B A G X A A A
B P A X B A A
-------------
D H A R H R B
```

Find the digits for which the letters stand.

55. Addition: Four Numbers

Below is an addition sum with letters substituted for digits. The same letter stands for the same digit wherever it appears, and different letters stand for different digits:

```
        E
      K X
    Y K E
    Y Y K X
  E E P D K
```

Write the sum out with numbers substituted for letters.

56. A Multiplication

In the multiplication sum below the digits have been replaced by letters. The same letter stands for the same digit wherever it appears, and different letters stand for different digits.

```
  B X Y D X B T
              T
  _____
B M X R Y P D T
```

Write the sum out with letters replaced by digits.

Part VIII

?

Some Letters or
Digits Missing

57. Division: Some Missing Figures

The following long division sum, with most of the figures missing, comes out exactly.

```
            _  _
 _  _ ) _  _  _  7
       _  5
       _  _
       _  _
```

Find the missing figures.

58. A Division Sum

```
                    2  -  -
      -  - ) -  -  -  -  -  7
               -  -
             4  -  -
             -  -  -
                -  -
                -  -
```

Find the missing digits.

59. Long Division

In the following division sum most of the digits are missing but some are replaced by letters. The same letter stands for the same digit wherever it appears.

```
              _  _  _
  _  m ) _  _  _  _
         _  p
         p  m  _
         _  _  _
            x  _  x
            p  _  _
               p  _
```

Find the correct sum.

60. Long Multiplication

In the following multiplication sum letters have been substituted for most of the digits.

```
        _  _  _  _
              x  g
    ─────────────────
       y  b  h  m  g
       g  y  x  g
    ─────────────────
    x  d  x  d  p  g
    ─────────────────
```

Write out the whole multiplication sum.

61. Division: Some Letters for Digits, Some Missing

In the following division sum some of the digits are missing and some are replaced by letters. The same letter stands for the same digit wherever it appears.

```
              y  -  -
      ─────────────────
- y ) -  -  -  -
        -  b
      ─────────
        -  -  -
      d  -  -
      ─────────
         d  -  -
         e  -  -
      ─────────
            d  e
```

Find the correct sum.

Part IX

?

Verse and Miscellaneous

62. Alphabetical Chairs

Alf, Bert, Charlie, Duggie, Ernie and Fred are sitting round a circular table. No two men, the initial letters of whose names are next to each other in the alphabet, are next to each other at the table.

Duggie is sitting opposite to Alf, who has Ernie's sister's husband on his right.

Draw a sketch showing how everyone is sitting relative to Duggie.

63. Football in Verse

Football, football, that's this story,
Five teams will play each other once.
Some will get success and glory,
And one of them may be the dunce.
  Let us call them A, B, C
  And D, and then (you've guessed it!) E.
E played them all, but won no fame
For they did not win a game.
  And A and B, why yes, and D
  Had goals against, 3, 4 and 3;
  In each case respectively.
C only won a single match, no more;
No draws, no goals against in any game.
A's matches played, B's wins and B's goals for,
Were all an even number and the same.
  This only leaves us this to say
  A scored three goals, C scored four
  And D, the fourth team, had one draw.

Find the score in each match.

64. 'What a way to end this song!'

Five teams, this time, will play each other once.
But when, ah when? We aren't in fact told much.
How many matches have they won? Who knows?
That column, I'm afraid, is quite, quite blank;
Unlike the one which tells us about points.
A, C and D got two points each, while B
And then the fifth team, E, got five and one.
A and D each drew a couple while B
And E drew one. No games were lost by B or A;
But C and E lost two and three.
Why, yes, of course, respectively.
Having changed the rhythm once, I'll do it twice and say,
Matches played by A, B, E were two and three and four.
E had thirteen goals against while D had seven for.
Five, two, four were goals scored for by B and C and E.
While seven and two for A and C were goals they had against.
   Oh! by the way, I ought to say
   A win is worth two points.
   And if the match is drawn, why, then
   The points are drawn as well.
   And now I must confess,
   And it makes me very sad.
   Of these figures *one* is wrong,
   What a way to end this song!

Find the score in each match.

65. Two, Three, Four, Six

In this long division sum, I fear,
Most of the figures simply are not there.
  Two and Three and Four and Six,
  One of these is wrong. But which?
Three and Six and Four and Two,
Do you think that's much too few?
  Why don't I give you rather more,
  Than Six and Two and Three and Four?
To give you four, you will agree,
Is better than to give you three.
  Look at the pattern if you wish,
  All the figures look like this:

```
            -  4
   -  6 ) -  -  -
            -  -
            -  2  -
            -  -  3
```

Which figure was wrong? Find the correct division sum.

66. Oh! Uncle, oh! Bungle

Oh! Uncle, oh! Bungle, you've done it once more!
I asked you to try, and indeed to make sure
That there was no mistake. But I'm sorry to see
A mistake there has been, as you'll surely agree.
    But I'd better explain what this puzzle's about.
    Football's the answer, with some facts left out.
    And we'll try to forget that my Uncle's a dunce!
    Four soccer teams have to play each other once.
A and B and D had one, four, six goals for
Respectively. And A had one against, no more.
B had five against while D had only three.
Two matches played, two matches drawn, that's C.
    B lost two games and got no points, while A
    And D got three and five points each, but say –
    And say again, and put it on your list,
    Of the figures, *one* is wrong. And then, a final twist.
        It's for you to discover
        If they've all played each other.

(2 points are given for a win, and 1 point to each side in a drawn match.)

Find the score in each match.

67. Logic Lane

I asked my love what number her domain,
> in Logic Lane.
'Prove,' she replied, 'how warm your adoration
> by calculation;
Not only sums, but also do some thinking,
> thereby linking,
The clarity and force of intuition,
> with my position.'

So meekly I replied:
  'Adored one, let me have the factual data
  And we'll discuss the metaphysics later.'
Swift to the point she plunged:
'My dwelling has three figures, different all.
And like, I hope, your love for me they rise.
Has it got factors? Yes, two different ones;
Prime numbers both, greater than ten plus three.
Add the digits and you then will find
The sum is in the upper half of what you have in mind.'

What is the number of my loved one's house?

68. 'Do you add or subtract? That's for you to discover'

Letters stand for digits, but who knows what stands for what?
  You've got to use your wits and thus decide.
Different letters, different digits, for I've got,
  As maker of the puzzle, proper pride.
E and R and I and C, that makes a name,
  And the second line is similar,
  With two letters familiar
For E and R read I and P, and then the same.
With an R and a V and a couple of E's,
  The sum is completed; but please
Do you add or subtract? That's for you to discover
If the one can't be done, then it must be the other.

```
E  R  I  C
I  P  I  C
---------
R  V  E  E
```

Do you add or subtract? Write the sum out with numbers substituted for letters.

69. Sons and Grandsons

Many years have passed since I first started my factory, and it is pleasant to be able to record that at least some of my original employees are now grandfathers. Alf, Bert, Charlie, Duggie and Ernie have between them six sons who are called:
Peter,
Quentin,
Roland,
Samuel,
Tristan,
Uvedale.
And these six sons have between them seven sons who are called:
Ivan,
Justin,
Kenneth,
Larry,
Michael,
Nicholas,
Orlando.
You are told that Bert, Charlie and Duggie's sons, if they had any, were Peter, Quentin and Tristan; that Alf, Bert, Charlie and Ernie's sons, if they had any, were Peter, Roland, Samuel and Uvedale; that Alf, Bert, and Duggie's sons, if they had any, were Quentin, Roland, Samuel and Tristan; that Bert, Duggie and Ernie's sons, if they had any, were Quentin, Tristan and Uvedale.

You are also told that Quentin, Roland and Tristan's sons, if they had any, were Ivan, Nicholas and Orlando; that Peter, Roland, Samuel and Uvedale's sons, if they had any, were Justin, Kenneth, Larry and Michael; that Quentin, Tristan and Uvedale's sons, if they had any, were Ivan, Kenneth, Nicholas and Orlando; that Peter, Samuel and Tristan's sons, if they had any, were Justin, Larry, Michael and Orlando.

One of Alf, Bert, Charlie, Duggie and Ernie has no sons. *Which one?* And one of Peter, Quentin, Roland, Samuel, Tristan and

Uvedale has no sons. *Which one?* (No one has more than two sons.)

Who is Kenneth's grandfather and who is Orlando's grandfather?

70. Add Twice

Below are two addition sums with letters substituted for digits. The same letter stands for the same digit wherever it appears in either sum, and different letters stand for different digits.

(i)
$$
\begin{array}{ccccc}
h & m & p & d & m \\
b & h & p & h & m \\
\hline
r & c & d & h & a
\end{array}
$$

(ii)
$$
\begin{array}{cccc}
r & b & a & d \\
p & q & h & d \\
\hline
a & a & d & d
\end{array}
$$

Find the digits for which the letters stand.

71. The Professor and the Islands of the World

One thing I can say about my boss – Professor Knowall – is that he is always prepared to move with the times (as a matter of fact there are a lot of other things I could say, too, but perhaps not now). He also likes moving around and is particularly keen on islands. Perhaps this is because he thinks of an island as a new idea and one can see what he means.

His latest visit was to the Island of Near Perfection, where the inhabitants try like mad to improve everything in every way every day and will, of course, or so they say, drop the 'Near' one of these days.

I am rather keen on football myself and was able to explain to the Professor the new ideas that were being put forward. It has become increasingly obvious over the years that goals are what the public like and the greatest happiness of the greatest number is what they are after on this island. In order that they shall score more often, the size of the goals has been greatly increased and the game now lasts rather longer.

The Professor, of course, got the point straight away.

'My dear Sergeant Simple,' he said, 'this will mean more and better puzzles. The guesser will no longer be successful and there will be much more success now for those who really use the minds which God has given them. Bring me a bit of paper, my dear Simple,' he said, 'and we will produce a nice easy puzzle straight away.'

And, as always, he was as good as his word. Here is the puzzle in which three teams, A, B and C, are eventually going to play each other once (or perhaps they have already done so). The information given seemed to me to be both very odd and also very insufficient, but perhaps my readers might like to look at it and see what they think.

Here it is:

	PLAYED	WON	LOST	DRAWN	GOALS FOR	GOALS AGAINST
A					28	22
B					26	
C				I	24	23

Find the score in each match.

Solutions

1. Greater or Lesser

(i) 4 down is the square of an odd number, ∴ it must be 25, 49 or 81. But from 5 across the second figure of 4 down cannot be 9 or 1.
∴ 4 down must be 25.

(ii) 3 across is a multiple of 4 down (i.e. of 25), ∴ since there are no o's it must end in 5.
∴ 3 across is _25.
∴ 1 across is (_25) multiplied by 3.
∴ 3 across must start with 1, 2 or 3, and 1 across must start with 3, 6 or 9.

But the sum of the digits of 1 down is 18, and if 3 across started with 1 or 2 and 1 across started with 3 or 6, the third figure of 1 down would be 10 or more.
∴ 3 across must be 325, and 1 across must be 975, and 1 down is 936.

(iii) The third figure of 5 across must be less than 5. And since 2 down is a multiple of 3, ∴ it must be 552.

Complete Solution

I		2
9	7	5
3 3	**4** 2	5
5 6	5	2

2. A Cross Number (3 by 3)

The digits of 1 across must be *both* even or *both* odd, since their sum is 10. ∴ 1 across cannot be the same as 3 down reversed (for from 3 across and *4* across the first digit of *3* down is even and the second odd).

3 across and *4* across cannot be the same when reversed, for all the digits of *3* across are even, and all the digits of *4* across are odd.

If *1* down were the same as *3* across reversed, then all the digits of *1* down would be even. But we know that the third digit of *1* down is odd (see *4* across). Similarly, *1* down cannot be the same as *4* across reversed, and *2* down cannot be the same as either *3* across or *4* across reversed.

.˙. the only numbers which can be the same when reversed are *1* down and *2* down.

.˙. since all digits of *4* across are odd, the digits of *1* across must both be odd.

.˙. they must be 1 and 9, *or* 3 and 7, *or* 5 and 5. But *not* 5 and 5 (see *4* across); and not 1 and 9, for the last digits of *4* across would then be 1 and 9 or 9 and 1, neither of which is possible.

.˙. *1* across must be 37 (not 73, see *4* across).

.˙. first digit of *4* across can only be 9.

Since *2* down (7_3) is a multiple of *3* down (_9) .˙. *2* down must be *3* down multiplied by 7, 17, 27, etc.

Suppose 3 down were 89 (the first digit must be even). Then 89 × 7 = 623. No. .˙. first digit is not 8. *Suppose 3 down were 49.* Then 49 × 7 is too small, 49 × 17 = 833, and this is too large. .˙. *3* down can only be 29. And 29 × 27 = 783. .˙. *2* down is 783 and *1* down is 387.

Complete Solution

	1 3	*2* 7
3 2	8	8
4 9	7	3

3. A Cross Number (5 by 5)

(i) 6 across is a square. It cannot end in 1 or 9 (see 4 down), ∴ it must be 16, 25, 36 or 64. ∴ the second figure of 4 down is 4, 5 or 6; ∴ the last figure of 4 down is 5 or more; i.e. the first figure of 8 across is 5 or more, ∴ the second figure of 3 down is 5 or more. ∴ 6 across must be 64.
  ∴ the first figure of 8 across is 6, ∴ 8 across is 65, ∴ 3 down is 56. And the first figure of 4 down is 1 or 3 (see 1 across).

(ii) 12 across must end in 2 (see 12 across and 9 down). ∴ it starts with 2. And the second figure of 9 down must be 3 or 4.

(iii) *Consider 1 down.* Since there are no 0's, it must end in 5 (a multiple of 65); and since it starts with an odd number it must be 195 or 325 (*not* 585, for 3 down starts with 5).
  5 across cannot start with 9, for the second figure would then be 1, but this is not possible (see 2 down). ∴ 1 down is 325, and 5 across is 28, and 2 down must start with 9. And since 3 is the first figure of 1 across, ∴ 4 down starts with 1, and 1 across ends in 7.

(iv) *Consider 7 across.* It must be 53 or 54, but not 53 (see 2 down), ∴ 54; ∴ 9 down is 542, and 2 down is 98432. 10 across must be 36, ∴ 11 down is 66, and 12 across is 2662.

Complete Solution

¹ 3	² 9	³ 5	⁴ 1	7
⁵ 2	8	⁶ 6	4	▨
⁷ 5	4	▨	⁸ 6	⁹ 5
▨	¹⁰ 3	¹¹ 6	▨	4
▨	¹² 2	6	6	2

113

4. A Cross Clue

We must look first for the incorrect clue.

Consider 5 across and 3 down. If both true then 5 across can only be 27 (not 64 for each digit of 3 down is odd). But 3 down cannot then be correct for, if it were, its third digit could not be more than 5.

∴ either 3 down or 5 across must be wrong; and all other clues must be correct.

1 down is a multiple of 17, and from *1* across and *4* across the first digit is odd and the second is even. ∴ it can only be 34.

4 across is a multiple of 37 (34 + 3). And the first figure is 4. The only possibilities are 407, 444 and 481. But there are no 0's, and the digits are all different. ∴ 4 across can only be 481.

∴ 3 down must be the incorrect clue, for each digit cannot be less than the one before. ∴ 5 across is correct and must be 27 or 64. But not 64, for there is no multiple of 34 which will do for 2 down. And 2 down can only be 782 (34 × 23). *1* across can only be 379.

Complete Solution

3 down is incorrect.

1 3	*2* 7	*3* 9
4 4	8	1
//////	*5* 2	7

5. Some Very Cross Numbers

1 8, 9	2	3
4 1, 2	5 1, 2	6 2
7 8, 9	8	9

(i) From 3–6 (6) must be 2, 4, 6 or 8. From 4–5–6 the sum of (4) (5) and (6) must be less than or equal to 5: ∴ since there are no 0's (6) must be 2 and (4) and (5) must each be 1 or 2. (Insert in diagram as shown.)

(ii) From 1–4–7 (1) + (4) + (7) = 19 or more. And since (4) is 1 or 2, ∴ (1) and (7) must each be 8 or 9. (Insert in diagram as shown.)

(iii) 1–2–3 must be a multiple of 3, 5 and 7; ∴ a multiple of 105. And since (1) is 8 or 9, ∴ it must be 105 × 8 or 105 × 9. But 105 × 8 (840) has a 0, ∴ 1–2–3 is 105 × 9, i.e. 945.

(iv) The sum of digits of 2–5–8 < 5 + 2, ∴ (8) must be 1 or 2.

(v) From 7–8–9 (9) must be > 8 or 9 + 1 or 2, ∴ (9) must be 9, ∴ (7) must be 8 and (8) must be 1.

(vi) 1–4–7 must be 928, ∴ 4–5–6 must be 212.

Complete Solution

1 9	2 4	3 5
4 2	5 1	6 2
7 8	8 1	9 9

6. Odds and Evens

(i) *Consider 7 down (a square).* The first digit must be even (6 across), and the second digit must be 1, 3, 7 or 9 (9 across). The only square which satisfies these conditions is 676 (the square of 26).

(ii) 3 down must end in 6 (see *11* across). *1* across is a multiple of 6 across, ∴ it is even. ∴ 3 down can only be 23456.

(iii) By considering 4 across we see that 4 down must start with 1, 5 or 7 (not 3 because 3 is the second figure of 3 down, and not 9 because there is no cube 9_ _). But not 1 (4 down would then be 125 and 57 is not a prime). And not 5 (4 down would then be 512 and 27 is not a prime). ∴ 4 down is 729 (cube of 9). ∴ 8 across is 45.

(iv) *Consider 5 down.* The first digit must be greater than 5, and since 4 down starts with 7, ∴ 5 down must be 95_.

(v) *1* down is a square, and it must end in 1 or 5 (see 4 across – these are the only two left), ∴ it is 81 or 25.

If 25, then *1* across is 2_2 and this must be a multiple of 26. For a multiple of 26 to end in 2 we must multiply by 2 or 7: but $26 \times 7 = 182$; $26 \times 17 = 442$; $26 \times 2 = 52$; $26 \times 12 = 312$. ∴ *1* down is not 25.

But if *1* down is 81, then *1* across, 832, fulfils the conditions – and nothing else does.

∴ 4 across must be 71539.

(vi) *10* down is a multiple or a factor of 2 down. There are no two-digit factors of 35, except itself. The only two-digit multiples of 35 are itself and 70, but there are no 0's. ∴ *10* down must be 35.

(vii) From 5 down and *10* across it follows that *10* across is 351, and 5 down 951.

Complete Solution

	1 8	2 3	3 2	
4 7	1	5	3	5 9
6 2	7 6		8 4	5
9 9	7	10 3	5	1
	11 6	5	6	

7. A Crosser Number

(i) We must first try to find the incorrect clue: 9 across must be 19,38,57,76 or 95. In every case the first digit is *odd*. But *8* down is *even*. ∴ *either 9 across or 8 down* is the incorrect clue. ∴ all the other clues are correct.

(ii) *Consider 3 down (a cube).* The first digit must be 6 or less (see *1* across), and the last digit must be even (see *6* across), ∴ 3 down must be the cube of 6 (216) or of 8 (512). In both cases the second digit must be 1, therefore *5* across is 81, and the fourth digit of 2 down is 8.

(iii) *Consider 10 across.* If 3 down is 512, then 'twice the square of the cube root' is $2 \times 8 \times 8 = 126$. But this has 3 figures, ∴ 3 down cannot be 512, ∴ it must be 216, and *10* across is $2 \times 6 \times 6 = 72$. ∴ the first digit of 2 down is 7, and *1* across is 9872.

(iv) The first digit of 7 down must be 2, 4 or 8 (see *6* across). But a perfect square cannot end in 2 or 8 (see *7* down), ∴ 7 down must be 46.

(v) *Consider 2 down and 8 across.* 2 down is 7, 8, 2 or 8, 8, 7, so that the digits add up to at least 32. 8 across is _ _68 so that digits can only add up to 32 if first two digits are both 9.∴ 8 across must be 9968 and the third digit of 2 down must be 2; ∴ the first digit of 6 across is 8.

(vi) *Consider 4 across.* The last two digits of 8 across are 68, which is 2 × 2 × 17.∴ 4 across must be 17.

(vii) *Consider 1 down.* The last digit must be 1.

(viii) 9 across must end in 9, 8, 7, 6 or 5, not 1 (see (i)).∴ 9 across is the incorrect clue. For 8 down to be correct the second figure could be 2, 4, 6 or 8.

Complete Solution

9 across is the incorrect clue.

	1 9	8	2 7	3 2
4 1	7		5 8	1
	6 8	7 4	2	6
8 9	9	6	8	
9 2,4,6 or 8	1		10 7	2

8. One Good Score

(i) Since A won 1 match and got 16 points, ∴ they must have lost 1 (o points) and won 1 on the first innings (6 points).

(ii) C got 21 points, and they can only have got this in 3 matches by getting 10, 6 and 5.

C v. A cannot have been drawn (A drew none), and it cannot have been a win for C on first innings (A lost none on first innings).

∴ C beat A (10 points for C, 0 for A).

(iii) With 6 points D cannot have drawn any, ∴ drawn match was B v. C (5 points for each).

∴ C v. D (6 points) was a win for C on first innings.

∴ since D got 2 points from match v. C, they must have got their other 4 points from a defeat on the first innings v. A, and a defeat on the first innings v. B. And the match that A won was v. B.

Complete Solution

A v. B : A won
A v. C: A lost
A v. D: A won on first innings
B v. C: tie
B v. D: B won on first innings
C v. D: C won on first innings

9. One Winner

(i) *Consider B's points (11).* The only way of getting this from 3 matches is 6, 5, 0.

Consider A's points (10). This could be 10, 0, 0; or 5, 5, 0; or 2, 2, 6. But the first two of those would mean that at least 2 matches were won (for if A gets 0 someone else gets 10).

∴ A's points must be 2, 2, 6.

∴ B v. A must have been a win on the first innings for B.

(ii) If D had drawn a match (5 points), they would then have had to get 9 points from 2 matches. But this is not possible. ∴ B's drawn match was v. C.

∴ B v. D was a win for D.

∴. D got 2 points from each of their matches v. A and v. C.
And C v. A was a win on the first innings for C.

Complete Solution

A v. B : B won on first innings
A v. C: C won on first innings
A v. D: A won on first innings
B v. C: tie
B v. D: D won
C v. D: C won on first innings

10. Cricket: Four Teams

(i) If A, B, C and D had all played each other, 6 matches would
have been played and the greatest possible number of points
would have been 60.
    In fact A, B and C got 56 points (18 + 17 + 21), so all 6
matches were played.

(ii) If all matches had been won or tied each team's points
would have been a multiple of 5. But none of A, B and C's
points are multiples of 5. ∴. each of these teams must have
played at least 1 match in which one team got 6 and the other
2. ∴. there must have been at least 2 such matches.
    But there cannot have been more than 2, for A, B and C's 56
points are only 4 less than the possible total of 60, and for a
match that is won on the first innings 8 points (6 + 2) are given
instead of 10. ∴. D got no points and lost all 3 matches. B's 17
points must have been 10 for a win, 5 for a tie and 2 for a defeat
on the first innings. C's 21 points must have been 10 for a win,
5 for a tie, and 6 for a win on the first innings. A's 18 points
must have been 10 for a win, 6 for a win on the first innings, 2
for a defeat on the first innings.

(iii) ∴. B v. C was a tie. A v. B was a win on the first innings for
A. C v. A was a win on the first innings for C.

Complete Solution

A v. B : A won on first innings
A v. C: C won on first innings
A v. D: A won
B v. C: tie
B v. D: B won
C v. D: C won

11. Latest Results

(i) If they have all played each other, 6 matches would have been played and the total of points would be at least 48 (if all matches were drawn) and at most 60 (if all matches were finished).

In fact the total of points (21 + 10 + 9 + 6) is 46 . ˙. 1 match had not been played.

(ii) Since A has got more than 20 points, they have played 3. And they must have won 1 (10 points), won 1 on their first innings (6 points) and tied 1 (5 points).

C has 9 points. There is no way in which they could have got this from 2 matches, . ˙. they must have played 3. 1 was drawn, and 2 were lost on the first innings.

. ˙. the match that has not been played is B v. D.

(iii) . ˙. since D only played 2, they must have got their 6 points from a win on the first innings. But this cannot have been v. A (who lost none on the first innings), . ˙. D v. A was a win for A. And D v. C was a win on the first innings for D.

(iv) B's 10 points cannot have been for a win for neither A nor C lost a match.

. ˙. both B's matches were tied.

. ˙. A v. C was a win on the first innings for A.

Complete Solution

A v. B : tie
A v. C: A won on first innings

A v. D: A won
B v. C: tie
C v. D: D won on first innings

12. And Where Was B?

If all 6 matches were won, 60 points would have been obtained. In fact the points were at least 53 (18 + 17 + 10 + 8), and at most 55 (18 + 17 + 10 + 10). But, as the total of the points must be even, it can only be 54, and since this is 6 points less than the possible figure, 3 matches must have been drawn (10 points if match is finished, but only 8 points (6 + 2) if it is not).

D got 10 points. These must *either* have been 0, 5, 5 *or* 2, 2, 6.

Suppose 2, 2, 6. Then all the other 3 matches would have been won or drawn.

∴ 2 of A's 3 matches would have been won or drawn. But with 18 points, A cannot have won 2 and there is no way of their drawing 1. (18 − 5 = 13, and it is not possible to get 13 points from 2 matches.)

∴ D's points must have been 0, 5, 5. And since A cannot have drawn 1, we must have:

	A	B	C	D	POINTS
A	✕			10	18
B		✕		5	more than 7, less than 11
C			✕	5	17
D	0	5	5	✕	10

∴ the other 3 matches were not finished, and one side must have got 6 points and the other 2 points.

Since C got 17 points, C v. A and C v. B were both won on the first innings by C. And since A got 18 points, A v. B was won on the first innings by A.

Complete Solution

A v. B : A won on first innings
A v. C: A lost on first innings
A v. D: A won
B v. C: B lost on first innings
B v. D: tie
C v. D: tie

13. Can You Draw Any Conclusions?

Consider C. They won none, but they got 2 points; ∴ they must have drawn 2. But not v. D who got no points. ∴ C v. A and C v. B were both drawn, and C did not play D.

Consider B. They got 5 points, and B v. C was a draw; ∴ B v. A and B v. D were both won by B.

Consider A. They drew v. C and lost v. B, and they got 4 goals for and 4 against. ∴ they must have won v. D. (Otherwise goals against would be more than goals for.)

We now know who played whom and the result of each match.

C drew the 2 matches they played, ∴ scored 4 goals for. And since the total of goals for must be the same as total of goals against, ∴ B had 2 goals against. Since B only had 2 goals against, ∴ B v. C must be 2–2, 1–1 or 0–0. But not 0–0, for C v. A would then be 4–4, and this is not possible, for A scored at least 1 of their 4 goals in beating D.

If B was 1–1, then C v. A would be 3–3, A v. B would be 0–1 and A v. D would be 1–0. From B's goals for and against, B v. D would be 3–1. But this is not possible, for D's goals for and against should be 0–3 (*not* 1–4).

∴ our assumption was wrong, and B v. C cannot be 1–1.

∴ B v. C can only be 2–2.

∴ C v. A is 2–2.

B only had 2 goals against, ∴ B v. A is ?–0. And since A had 4 goals for, ∴ A v. D is 2–?. And since D had *no* goals for, A v. D is 2–0. ∴ A v. B is 0–2. ∴ B v. D is 1–0.

A v. B : 0–2
A v. C: 2–2
A v. D: 2–0
B v. C: 2–2
B v. D: 1–0

14. C's Loss

(i) B played 4, ∴ they have played everybody. ∴ C's match was v. B, and the score was 1–3.

(ii) D had 2 goals for and 1 against, and lost 0 and drew 0.
∴ they played only 1 match and the score was 2–1 – and this match must have been against B. (If they had played 2 matches they could not have won them both, for goals for is only 1 greater than goals against.)

(iii) A drew 0, ∴ they must have played 2 matches and won 1 by 1–0, and lost the other by 0–1.
One of these matches must have been v. B and since B lost only 1 match (v. D, 1–2), B v. A was 1–0.

(iv) We now have:

	A	B	C	D	E
A		L 0–1			
B	W 1–0		W 3–1	L 1–2	
C		L 1–3			
D		L 2–1			
E					

A's other match must have been v. E, and score was 1–0. And since B played everyone they played E, and since B had 9 goals for and 5 against, ∴ score in B v. E was 4–2.

Complete Solution

A v. B : 0–1
A v. E : 1–0
B v. C: 3–1
B v. D: 1–2
B v. E : 4–2

15. Four Teams: Letters for Digits, One Letter Wrong

We must first try to find out more about the letter that is wrong.

Consider C. If they won x and got x points, x could only be 0 (if they won one they would have got at least 2 points). But if x were 0, then B's played would be wrong, for B had goals for and against.

∴ *either* C's win (x) *or* C's points (x) *or* B's played (x) must be wrong. ∴ all else is correct.

Suppose that the correct x were 0. Then B would have got no points. But this is not possible, for B had y goals for and y against, and must therefore have got at least 1 point. ∴ x is not 0.

And y (C's played) cannot be 0, for C had goals for and against.

y, t, p and the correct x must be 0, 1, 2 and 3 in some order, for the number of matches played, won, lost and drawn by any team cannot be more than 3. And since neither x nor y can be 0, ∴ *either t or p* must be.

And since x is 1 or more, ∴ *either* C's wins *or* C's points must be wrong. ∴ B's played is correct. Suppose B's played (x) were 1, then B would have got x points and would therefore have drawn 1. But this is not possible, for B's draws (p) would then also be 1. ∴ x cannot be 1.

Suppose x were 3. Then B would have got 3 points (x). ∴ they would have won 1 and drawn 1 (they could not have drawn 3, for p

and x cannot both be 3). But this is not possible with y goals for and y against.

∴ x can only be 2.

∴ B played 2 and got 2 points. If they had drawn 2, p would also be 2. And since it is not, B must have won 1 and lost 1 and p must be 0.

If t (A's draws) were 3, then A would have drawn against all the other three. But we know that B drew none. ∴ t must be 1, ∴ y must be 3.

D lost 2 and got no points, ∴ they can only have played 2.

C played 3 and B played 2, ∴ A must have played 1 or 3 (to make total of matches played even). A drew 1 and got 1 point and had h goals for and h against. ∴ if they had played 3 they would have won 1 and lost 1 as well and would therefore have got 3 points. ∴ A can only have played one.

Since they had h goals for and h against, it was a draw. And since C played all the others, A v. C was a draw. D got no points, ∴ they lost both their matches.

And we have:

	A	B	C	D	GOALS FOR	GOALS AGAINST	POINTS
A			Dr		h	h	1
B			L	W	3	3	2
C	Dr	W		W	g	h	
D		L	L		0	h	0

We can now see that C got 5 points, ∴ the letter that was wrong was C's points (x).

It should be 5, not 2 (x).

Since p (D's goals for) is 0, ∴ D v. B and D v. C are both 0–?.

And since B had 3 goals against, B v. C is ?–3.

A v. C is h–h, and C had h goals against, ∴ C v. B and C v. D were both ?–0. ∴ B v. C is 0–3 and B v. D is 3–0.

We know that h (A's goals for and against) is more than 3.

Suppose it is 5. Then, since D had h goals against, and D v. B was 0–3, ∴ D v. C would be 0–2. But C's goals for (g) would then be 10

(5 + 3 + 2), which is not possible. And if h were more than 5 the situation would be worse; ∴ h must be 4.
∴ A v. C is 4–4, and since h is 4, D v. C is 0–1, and g is 8.

Complete Solution

The letter that is wrong is C's points, it should be 5, not 2 (x).

A v. C: 4–4
B v. C: 0–3
B v. D: 3–0
C v. D: 1–0

16. Four Teams

D won 2, ∴ they must have lost 1 to make goals for equal to goals against. ∴ D played them all and drew none.

B got no points, ∴ they did not draw v. C. ∴ C's drawn match must have been v. A. And since B got no points, D v. B was won by D.

A only played 2, ∴ A did not play B, as he had to play D. B played 1 or 2 matches. Suppose B played 1 (v. D), then score in this match was ?–4. But D only scored 3 goals, ∴ B v. D cannot have been B's only match. ∴ B played C, and this match was won by C, for B got no points. Total of goals against must be equal to total of goals for, ∴ A and B between them scored 5 goals (5 + 7 + 3 = 3 + 4 + 5 + 3). B had 4 goals against in 2 matches which they lost, ∴ B could not have scored more than 2 goals (and might have scored less). ∴ A scored at least 3 goals. But if A scored 3 goals, then *both* their matches would have been drawn. But this is not possible, for there is no one for their second drawn match to be against. ∴ A must have scored more than 3 goals.

∴ A v. D was won by A.

∴ since D won 2 matches, D v. C was won by D.

Consider C. Since they drew 1, lost 1 and won 1, and had 7 goals for and 5 against, ∴ the match which they won (v. B) was won by a margin of at least 3 goals. But B lost both their matches (v. C and v. D) and had 4 goals against. ∴ B v. C must have been ?–3, and B v. D must have been ?–1. And since we know that D won this game it must have been 0–1. And since C beat B by a margin of at least 3 goals, ∴ C v. B was 3–0. D scored 3 goals, 1 against B, at least 1 against C (D won this match), ∴ either 1 or 0 against A.

Suppose 1 (so that A v. D was ?–1). Then A v. C would have been 2–2, and C would have scored 2 goals v. D (7 goals altogether). And since D won this match the score would have been at least 3–2. But this is not possible for D only scored 3 goals altogether (and 1 was against B).

∴ D scored *no* goals v. A, ∴ 2 goals v. C.

∴ since A had 3 goals against, A v. C was 3–3.

∴ C scored 1 goal v. D (7 altogether). ∴ C v. D was 1–2, and D v. A was 0–2.

Complete Solution

A v. C: 3–3
A v. D: 2–0
B v. C: 0–3
B v. D: 0–1
C v. D: 1–2

17. You Can't Do One Without the Other

In (i), since only 3 matches can be played, the total of points cannot be more than 6 (2 points for each match).

From (ii), d, x and t can none of them be 0. ∴ they can only be 1, 2 and 3 ($1 + 2 + 3 = 6$). And from second line down in (ii), t must be 3, and d and x must be 1 and 2 or 2 and 1.

C's wins (p) must be 0, 1 or 2. But it cannot be 1 or 2. ∴ $p = 0$. And since C won none, x must be 1 (not 2, for if C drew both their matches, their goals for would be the same as their goals against). ∴ d must be 2. And from (ii) $m = 5$.

B got 3 points and C got 1 point, ∴ either B v. C was a draw, or B and C both drew against A. But this is not possible, for A's goals for are not the same as their goals against. ∴ B v. C was 0–0 and C v. A was 0–3.

B v. A was 5 (m)–y. And since we know that B won this game, y can only be 4. . . . B v. A was 5–4.

Complete Solution

(i)   A v. B : 4–5
     A v. C: 3–0
     B v. C: 0–0

(ii)   3   2
     2   1
     —————
     5   3

18. Return Matches

(i) C played 3. . . . A and B once each and one of them twice.
(ii) A won 1 and drew 1 and had same number of goals for as against, . . . A played at least 3. . . . B played at least 1 v. A and 1 v. C.
(iii) If B played 2 matches against either A or C they would have scored at least 1 goal (see conditions). But B scored no goals. . . . B played only 1 v. A and 1 v. C. . . . C's third game was v. A, not B. . . . A played 2 v. C and 1 v. B.
(iv) Matches played are thus:

	A	B	C
A	✕	✓ / ✕	✓ / ✓
B	✓ / ✕	✕	✓ / ✕
C	✓ / ✓	✓ / ✕	✕

(v) We know that B drew none, . . . B lost both their matches, and since 'in no 2 matches were the scores exactly the same', the scores must have been 0–1 and 0–3.

A must have scored at least 1 goal in the 2 matches v. C (see conditions) and, since A only scored 3 goals altogether, .˙. A v. B was 1–0, B v. C was 0–3, and A drew one of their matches v. C and lost the other. (We now know the result of all the matches.)

(vi) If A's drawn match v. C was 1–1, then according to conditions A's lost match could be 0–2 or 2–3, but either of these would make A's goals for and against wrong. If A's drawn match was 2–2, then A's other match v. C would be 0–1. But this is the same as A v. B.
.˙. A's drawn match v. C must be 0–0.
.˙. A's lost match v. C was 2–3.

Complete Solution

A v. B : 1–0
B v. C: 0–3
A v. C: 0–0
A v. C: 2–3

19. D's Gain

(i) *Consider A.* They won none and drew none, but their goals against were only 1 more than their goals for. .˙. they can only have lost 1, .˙. they played 1 and the score was 3–4.

(ii) C only had 2 goals against. .˙. they did not play A, who scored 3 goals in their match.
.˙. C played all the other three teams.
And, since D only played 1 match, the score was 2–1, .˙. D v. C was 2–1. And A did not play D.
And, since A had 4 goals against in their one match, and E only scored 3 goals, .˙. A did not play E. .˙. A v. B was 3–4. And D's match v. C was D's only match.

(iii) C only had 2 goals against. .˙. B and E both scored 0 (D scored 2). B drew none, .˙. C v. E was a draw (0–0).

And, since E's goals were 3–2, ∴ E played B and score was 3–2.

And, since C scored 4 goals, ∴ C v. B was 3–0.

Complete Solution

A v. B : 3–4
B v. C: 0–3
B v. E : 2–3
C v. D: 1–2
C v. E : 0–0

20. Football and Addition: Letters for Digits

In (ii) g must be even (s + s).

∴ since in (i) g must be 0, 1, 2 or 3 (for no one can play more than 3 games), g must be 0 or 2. But not 0, for D got some points, ∴ g = 2. ∴ m (A's played) must be 1 or 3 (not 0, for A had goals against).

Suppose m were 1. Then D's points would be 1, ∴ D would have drawn 1. ∴ y (D's drawn) should be 1, but m and y cannot both be 1. ∴ m cannot be 1, and m = 3.

∴ since D got 3 points they must have drawn 1, ∴ y = 1.

In (ii), since g = 2, and s cannot be 1, ∴ s = 6. And since x cannot be 0, 1, 2 or 3, ∴ x must be 4. And since there is 1 to carry from second line down, p = 9.

A's points cannot be 1, 2, 3, 4 or 6. Since they cannot be more than 6 (3 matches), they must be 0 or 5.

The number of matches played cannot be more than 3 + 3 + 2 + 2, i.e. 10. But the points (if A got 5) would be 5 + 2 + 2 (at least) + 3, i.e. 12. But they should be the same, for each match appears twice and there are 2 points for each match.

∴ A's points cannot be 5 and must be 0. ∴ t = 0. Suppose B got their 2 points from 2 draws, then since their goals for are different from their goals against, they would have had to have played 3, and C would have played 2. ∴ (3 + 3 + 2 + 2), i.e. 10 matches would have been played.

But the points could not be more than $(0 + 2 + 3 + 3)$, i.e. 8. And this is not possible (see above). $\therefore$ B must have got their 2 points from a win, and D's drawn match was v. C.

$\therefore$ The points were $(0 + 2 + 3 + 3)$, i.e. 8. $\therefore$ since C played 2 matches, B only played 1, making the played column $(3 + 1 + 2 + 2)$, i.e. 8.

$\therefore$ since B only played 1, it must have been v. A.

$\therefore$ B v. A was 6–3.

$\therefore$ B scored 6 goals against A and since A lost their matches v. C and v. D, C and D must each have scored at least 1, making a total of 8 goals against A. h (A's goals against) cannot be 9, for p is 9.

$\therefore$ h can only be 8. And A v. C and A v. D are both 0–1.

C had 4 (x) goals against. None of them were scored by A, $\therefore$ C v. D must have been 4–4.

Complete Solution

(i) A v. B : 3–6
    A v. C: 0–1
    A v. D: 0–1
    C v. D: 4–4

(ii)

4	6
4	6
9	2

21. Seven Teams

Consider C. They lost none, so they scored points for each match they played. They had 5 goals for and 5 against, $\therefore$ they must have got their 3 points from 3 draws (if they had won 1 and drawn 1, they would also have had to have lost 1 to make their goals for the same as their goals against).

$\therefore$ C played 3 and drew 3.

A only played 1 and had 4 goals for and 2 against; $\therefore$ they drew none. And since B got no points, they also drew none. E played 2 and got 4 points, ∴ they drew none. ∴ C's drawn match must have been v. D, v. F and v. G.

G only played 1 match (v. C) and this was a draw. ∴ G v. C was 3–3 (G had 3 goals against). F played 2 and drew 1 and had no goals for and 8 against, ∴ F's drawn match (v. C) was 0–0, and they lost one (0–8). The score in C's third drawn match (v. D) was 2–2. Since none of A, B, C, E and G had as many as 8 goals for, F can only have lost to D. ∴ F v. D was 0–8; and F did not play anyone else.

The situation now is:

	A	B	C	D	E	F	G
A	╳		╳		╳	╳	╳
B		╳	╳			╳	╳
C	╳	╳	╳	Dr 2–2	╳	Dr 0–0	Dr 3–3
D			Dr 2–2	╳	W 8–0	╳	╳
E			╳		╳	╳	╳
F	╳	╳	Dr 0–0	L 0–8	╳	╳	╳
G	╳	╳	Dr 3–3	╳	╳	╳	╳

E only played 2 matches and since they got 4 points they won them both. A only played 1 match which they won. ∴ E cannot have played A. ∴ the 2 matches which E won can only have been v. B and v. D.

The match which A played can only have been v. B or v. D.

Suppose it was v. B. Then A did not play D, ∴ D's other *lost* match was v. B. But this is not possible, for B did not get any points. ∴ A's match must have been v. D (∴ A v. D was 4–2).

We know that D lost 2 matches (v. A and v. E); but since they have got their 3 points by winning against F and by drawing v. C, they cannot have played B, for B got no points.

∴ B only played 1 match v. E and the score was 3–?. And since E had 3 goals against, E v. D was ?–0. And since D had 8 goals against, D v. E was 0–2. ∴ E v. B was 4–3.

133

Complete Solution

A v. D: 4–2
B v. E : 3–4
C v. D: 2–2
C v. F : 0–0
C v. G: 3–3
D v. E : 0–2
D v. F : 8–0

22. Everyone Scores

(i) All matches have been played except 1. ∴ 5 matches had been played and 50 points awarded for wins and draws. Total points are 68, ∴ 18 goals were scored. Each team had played at least 2 games and had therefore scored at least 2 goals. Bearing this in mind, it is easy to see that we have:

	FOR WIN OR DRAW	FOR GOALS
A	20	6
B	–	4
C	15	3
D	15	5

(ii) C won 1, and drew 1 and scored 3 goals. ∴ C only played 2 matches (2 goals needed for a win, 1 for a draw).
D won 1 and drew 1 and lost none. ∴ D only played 2 matches. ∴ C did not play D. ∴ C's and D's drawn matches must both have been v. A.

(iii) B played and lost to A, C and D. ∴ we know the result of each match. And since C only scored 3 goals, ∴ C v. A was 1–1, and C v. B was 2–1. And we have:

	A	B	C	D
A	✕	W	Dr 1–1	Dr
B	L	✕	L 1–2	L
C	Dr 1–1	W 2–1	✕	✕
D	Dr	W	✕	✕

(iv) A v. D must be 1–1 or 2–2 (not more than 5 goals in any match).

Suppose 1–1. Then since A scored 6 goals, $\therefore$. A v. B was 4–1; and since B scored 4 goals, $\therefore$. B v. D was 2–?.

Consider D's goals. 1 v. A, $\therefore$. 4 v. B; $\therefore$. B v. D was 2–4. But not more than 5 goals were scored in any match. $\therefore$. our assumption is wrong; $\therefore$. A v. D must have been 2–2.

(v) $\therefore$. A v. B was 3–1 or 2; and D v. B was 3–2 or 1. If A v. B was 3–1, then A has 4 goals against; and if D v. B was 3–2, then D has 4 goals against.

But no two sides have same number of goals against. $\therefore$. A v. B must be 3–2, and D v. B must be 3–1.

Complete Solution

A v. B : 3–2
A v. C: 1–1
A v. D: 2–2
B v. C: 1–2
B v. D: 1–3

23. Three Teams

(i) 3 matches were played, $\therefore$. 30 points were awarded for wins or draws. A can only have got 20 of them, beating both B and C, and B and C must have got 5 each (B v. C a draw).

(ii) B scored 3 goals (8 − 5), ∴ . B v. C was 2–2 or 1–1, and B v. A was 1–? or 2–?.

So we have:

	A	B	C	GOALS
A	✕	W −1 −2	W	5
B	L 1– 2–	✕	Dr 2–2 1–1	3
C	L	Dr 2–2 1–1	✕	4

(iii) *Suppose B v. C was 1–1.* Then B v. A was 2–?, and C v. A was 3–? (C scored 4 goals). ∴ . since A beat B and C, they must have scored at least 3 + 4 = 7 goals. But they only scored 5. ∴ . B v. C cannot have been 1–1, ∴ . B v. C was 2–2. ∴ . B v. A was 1–?, and C v. A was 2–?. ∴ . A v. B must have been 2–1, and A v. C 3–2.

Complete Solution

A v. B : 2–1
A v. C : 3–2
B v. C : 2–2

24. Three More Teams

(i) *Consider C.* They cannot have won or drawn a match. But they must have played A and B, for if they had only lost to *one* of them, the score in that match would have been 4–5, or more (but not more than 6 goals were scored in any match).

If C had scored 3 goals in either of these matches the score would have been 3–4, or more. ∴ . C scored 2 goals v. A, and 2 goals v. B.

(ii) *Consider B.* They beat C, but this cannot have been their only match, for they would have had to score 10 goals in it. . ˙. B also played A. But they cannot have won this match, for they could not then have scored any goals in either match.

And they cannot have lost it, for they cannot have scored more than 4 goals v. C (C scored 2) and the score in B v. A would have had to be at least 6–7. . ˙. B v. A was drawn.

(iii) B must have scored at least 3 goals to beat C. . ˙. B cannot have scored more than 2 goals v. A. Suppose A v. B was 1–1, then A v. C would be 5–2, which is not possible. . ˙. A v. B was 2–2. . ˙. A v. C was 4–2; and B v. C was 3–2.

Complete Solution

A v. B : 2–2
A v. C: 4–2
B v. C: 3–2

25. All Scores Different

6 matches were played and . ˙. 60 points were awarded for wins or draws.

A cannot have got more than 20 points for wins or draws, B cannot have got more than 25 points (not 30, for in that case they would have scored at least 6 points for their goals in winning 3 matches), C got no points for wins or draws and D could not have got more than 15. That makes (20 + 25 + 15), which is 60 and . ˙. that must have been what happened.

We now have:

	WINS OR DRAWS	GOALS
A	20	5
B	25	9
C	–	6
D	15	5

D got 15 points for wins or draws but they cannot have drawn 3 matches, for C got no points for wins or draws, . ˙. D won 1 and drew 1, B must have drawn 1 (25 points for wins or draws) and A might have drawn 2 (1 v. B and 1 v. D).

Suppose that the only drawn match was B v. D. Then A won 2 and lost 1. One of A's wins could have been 2–1, but the other one must be at least 3–?.. ˙. A scored 5 goals in these 2 matches, but they must also have scored at least 1 goal in their lost match v. B.. ˙. this is not possible, . ˙. A must have drawn 2 (v. B and v. D) and won 1 v. C. And the scores in these 3 matches can only have been 1–1, 2–1, and 2–2, though we do not yet know which of the drawn matches was 1–1 and which was 2–2.

We know that B won their matches against C and D, and that C lost all 3 matches.

Consider D. Since A v. C was 2–1, D must have scored at least 3 goals against C. But as they only scored 5 goals altogether, D v. A must have been 1–1 and D v. B must have been 1–?. ˙. A v. B must have been 2–2, and D v. C was 3–?.

B scored 9 goals, 2 against A, and . ˙. 7 against C and D. B v. D cannot be 2–1, for A v. C is 2–1, . ˙. B v. D is 3–1 or 4–1 and B must have scored 4 or 3 goals against C.

C scored 1 goal v. A . ˙. 5 v. B and v. D. But not more than 2 goals v. D, and . ˙. 3 v. B.. ˙. C v. D was 2–3, C v. B was 3–4 and B v. D was 3–1.

Complete Solution

    A v. B : 2–2
    A v. C: 2–1
    A v. D: 1–1
    B v. C: 4–3
    B v. D: 3–1
    C v. D: 2–3

26. The Professor's Versatility

(i) A's and C's figures (8) are the same. . ˙. if one of these figures was wrong it would *not* be possible to discover which. . ˙. they must both be right. . ˙. *either* B's *or* D's figure is wrong.

(ii) 6 matches were played. ∴ 60 points were awarded for wins or draws. A and C could only have got 10 of these between them. D can only have got 30 of these (even if they won all 3 of their matches). ∴ B must have got 20 for wins or draws. But B's figure is only 19, ∴ B's figure is wrong and all the rest are right.

(iii) ∴ D got 30 points for wins and 27 for goals. ∴ D's score in all 3 matches was 9–1. A's and B's situation must be the same, ∴ they drew against each other and scored 1 goal in each match. ∴ we have:

	A	B	C	D
A		L 1–	Dr 1–1	L 1–9
B	W –1		W –1	L 1–9
C	Dr 1–1	L 1–		L 1–9
D	W 9–1	W 9–1	W 9–1	

(iv) We cannot tell the scores in B v. A and B v. C, though we know that B won them both. The scores in each case could be anything from 2–1 to 9–1. ∴ B got 20 points for wins and at least 5 points for goals and might have got as many as 19 for goals. ∴ B's points are between 25 and 39.

Complete Solution

B's figure was wrong (it should be between 25 and 39 inclusive).

A v. B : 1–2 to 9
A v. C: 1–1
A v. D: 1–9
B v. C: 9 to 2–1
B v. D: 1–9
C v. D: 1–9

27. More Goals

(i) D got 5 points, and ∴ could not have won or drawn any, but must have played at least 2 matches. (If D had only played 1 which was lost, the score would have been at least 5–6, but this is not possible.) A got 11 points, ∴ could not have won 1 (10 + 2 points needed); but must have played at least 2 matches, and might have drawn 1.

∴ A cannot have played D.

(ii) B got 8 points, ∴ could not have won one, ∴ did not play D. ∴ D played C and E and lost both matches. C got 12 points and beat D, ∴ the score in this match was 2–1 and this was C's only game. ∴ A who 'played at least 2 matches' played B and E only.

The situation now is:

	A	B	C	D	E
A	✕	✓	✕	✕	✓
B	✓	✕	✕	✕	✕
C	✕	✕	✕	W 2–1	✕
D	✕	✕	L 1–2	✕	L
E	✓		✕	W	✕

(iii) D v. C was 1–2, ∴ D v. E was 4–5. We know that E played A and D. Suppose these were E's only 2 matches. Then the greatest possible points are (10 + 5)(v. D) and (10 + 8)(v. A), i.e. 33. But E got 43 points, ∴ E played B.

We now know who played whom.

(iv) *Consider A v. B.* Neither side can have won this match, ∴ it was a draw. And neither A nor B can have drawn another match. ∴ E beat A and E beat B.

We now know the result of every match.

(v) A got 5 points for a draw v. B, ∴ A scored 6 goals. But A cannot have scored more than 4 goals in the *lost* match v. E. .˙. A v. B was at least 2–2. B only got 3 goals; .˙. 1 v. E and only 2 v. A. .˙. A v. B was 2–2, .˙. A v. E was 4–5. .˙. E scored 3 goals v. B. .˙. E v. B was 3–1.

Complete Solution

A v. B: 2–2
A v. E : 4–5
B v. E : 1–3
C v. D: 2–1
D v. E : 4–5

28. Goals Rewarded Rewarded

(i) 3 matches were played; .˙. there were 30 points for wins or draws.

A could not have got more than 10 points for wins or draws;
B could not have got more than 15 points for wins or draws;
C could not have got more than 5 points for wins or draws.
But this makes 30. .˙. we have:

	FOR WINS OR DRAWS	FOR GOALS AND BONUSES
A	10	6
B	15	5
C	5	6

(ii) *Consider B.* They must have drawn 1 (v. C) and won 1 (v. A). B v. C must have been at least 2–2 (there was a bonus in every match). B v. A must have been 2–1 (with 1 bonus for B).

We now have:

	A	B	C
A	✕	1–2	
B	2–1 ($b = 1$)	✕	2–2
C		2–2	✕

(b = bonus)

Since B did not get a bonus in B v. C, ∴ C did.

(iii) A v. C was won by A, and A got a bonus. But C did not. ∴ C scored 3 goals. ∴ A v. C must have been 4–3, with a bonus of 1 for A.

Complete Solution

A v. B: 1–2
A v. C: 4–3
B v. C: 2–2

29. Four Digits Divided by Two Digits

```
        2 3        (i)
  38 )2 1 9 2      (ii)
     2 1 0         (iii)
   ─────
       9 8         (iv)
       9 4         (v)
```

Consider (i). Since (iii) has three figures and (v) only has two, ∴ (i) should be 32 (remember that the figures in (i) are only 1 out). ∴ (iv) and (v) are twice the divisor, ∴ they must end in an even number. But not 8 or 4, for (iv) and (v) end in 8 and 4; and not 2 (for this is last figure in (ii) and would be brought down to (iv)). And the divisor cannot end in 0, for (iii) ends in 0.

. ˙. (iv) and (v) ends in 6. . ˙. the divisor ends in 3 or 8. But not 8, for this is the figure given. . ˙. the divisor ends in 3.

If the divisor were 33, then (iii) would be 99, but (iii) has three figures. And if the divisor were 53, then (iv) and (v) would have three figures (53 times 2 = 106).

. ˙. the divisor must be 43. And we know that (iii) and (v) are 43 times 3 and 2.

Add up from the bottom and we get:

Complete Solution

```
      3 2
43 ) 1 3 7 6
     1 2 9
       8 6
       8 6
```

30. Five Digits Divided by Two Digits

```
       3 4 2      (i)
39 ) 2 2 4 9 8    (ii)
     2 1 5        (iii)
       9 9        (iv)
       9 2        (v)
         8 6      (vi)
         7 5      (vii)
```

Since (iii) has three figures, and (v) only has two, . ˙. the first two figures of (i) must be 4 and 3 (*not* 3 and 4). Since (v) (two figures) is 3 times the divisor, . ˙. the divisor must be less than 34 (34 times 3 = 102).

And since the divisor times 4 is more than 100 (see (iii)), . ˙. the divisor is *more than* 25. And since the first figure given in the divisor is 3, . ˙. the divisor is 25, 26, 27, 28 or 29.

The last figure in (i) must be 1 or 3. If 3, then (vi) and (vii) would

143

start with 8 or 7 (3 × 25 = 75; 3 × 29 = 87). But (vi) and (vii) do start with 8 and 7. .˙. the last figure in (i) is not 3, and must be 1. .˙. (vi) and (vii) are the divisor. .˙. the divisor is not 26, nor 25, nor 28 (last figure in (ii)), nor 29 (the divisor is given as 39).

.˙. the divisor is 27. We know that (iii), (v) and (vii) are 27 × 4, 3 and 1.

Add up from the bottom and we get:

Complete Solution

```
          4 3 1
27 ) 1 1 6 3 7
     1 0 8
     ─────
         8 3
         8 1
         ───
           2 7
           2 7
           ───
```

31. Addition: Four Numbers

4	6	2	
2	6	4	
1	5	2	
2	0	1	
8	3	7	(iv)
(i)	(ii)	(iii)	

In (i) and (iii) the same four digits are added together, and since there is nothing to carry from (i), the sum of these digits is less than 10. None of them can be 0, for it would not make sense for any of the figures in (i) to be 0, .˙. the least they can be is (1 + 1 + 2 + 3), i.e. 7.

But, since 7 is the figure given in (iv), the correct digits must be 8 or 9. In (i) the same four digits are added together, but the result is different, .˙. there must be something to carry from (ii).

.˙. the last figure in (iv) (7) must be 8; and the first figure in (iv) (8) must be 9.

Suppose that 2 in (iii) stood for 3. Then we would have, at least, $(3 + 3 + 1 + 2)$, i.e. 9. But the last line down adds up to 8. .ˉ. 2 must stand for 1. .ˉ. 4 and 1 in (iii) can only stand for 2 and 4. .ˉ. 4 must stand for 2; and 1 must stand for 4.

And we have:

2	—	1	
1	—	2	
4	—	1	
1	—	4	
9	—	8	(iv)
(i)	(ii)	(iii)	

The digits left are 0, 3, 5, 6 and 7. And we know that there is 1 to carry from (ii). Suppose the second figure in (iv) were 0, then the figure above it would be at least $(3 + 3 + 6 + 5)$, i.e. 17 (too much).

The second figure in (iv) cannot be 3 (figure given).

Suppose the second figure in (iv) were 5. Then the first two digits of (ii) could not be 0 (for we could then only have $(0 + 0 + 6 + 7)$, i.e. 13, instead of 15).

If the first two digits were 3, the other two would have to add up to 9, but this is not possible, for we must not have another 3. The first two digits cannot be 5, for we are assuming that the last digit is 5. The first two digits cannot be 6, for this is the figure given. If the first two digits were 7, then the other two would have to add up to 1, which is not possible.

Suppose the second figure in (iv) were 6. Then the two digits above it (5 and 0) would have to stand for digits that add up to an *even* number. The digits left are 0, 3, 5 and 7. .ˉ. 0 cannot be one of the last two digits, for 0 + (3, 5 or 7) would be odd. .ˉ. (ii) would be at least 3 + 3 + 5 + 7, i.e. 18 – but this is too much ($(0 + 0 + 5 + 7)$ would be too small).

.ˉ. the second figure in (iv) can only be 7. The first two figures down cannot be 0 ($(0 + 0 + 6 + 7) = 13$), or 5 ($(5 + 5 + 0 + 7)$, *not* acceptable, for if 3 stands for 7 nothing else can). And the first two figures of (ii) cannot be 6 (figure given); nor 7, for if 3 stands for 7 nothing else can.

.ˉ. (ii) can only be:

```
3
3
6
5
─
7
```

(Note that since 5 cannot be the third figure down, for that is the figure given, $\therefore$ it must be the fourth.)

Complete Solution

```
2 3 1
1 3 2
4 6 1
1 5 4
─────
9 7 8
```

32. Five Digits Divided by Two Digits

It will be convenient to place a pattern beside the figures given in which the correct figures can be inserted as they are discovered. Thus:

```
        2 7 1      (i)          - - -
  1 3)2 0 6 3 4    (ii)      )- - - - -
      8 0          (iii)        - -
    ─────
      2 6 3        (iv)         - - -
      2 5 8        (v)          - - -
    ─────
          5 4      (vi)           - -
          4 6      (vii)          - -
        ─────
```

Since the digits in (i) are only 1 out, $\therefore$ the last figure in (i) must be 2. The divisor does not start with 1 (figure given). If the divisor was between 20 and 29, (vi) and (vii) would start with 4 or 5, but these are figures given, $\therefore$ the divisor does not start with 2.

Since the divisor $\times$ 2 has only two figures, $\therefore$ the divisor is less than 50. .˙. (iii) must be the divisor × 3 (for the divisor goes more than *once* into a three-figure number).

.˙. the divisor is 30, 31, 32 or 33 (not 34, for 34 times 3 = 102). But not 30 (see second figure of (iii)), and not 32, for 32 times 2 ends in 4, which is the second figure of (vi); and not 33, for 33 times 2 ends in 6, which is the second figure of (vii).

.˙. the divisor must be 31. .˙. (iii) is 93 and (vi) and (vii) are 62.

(v) must be 31 times 6 or 8, but not 8, for this is the last figure of (v);
.˙. (v) is 31 times 6, i.e. 186.

Add up from the bottom and we get:

Complete Solution

```
        3 6 2
 3 1 ) 1 1 2 2 2
        9 3
      1 9 2
      1 8 6
        6 2
        6 2
```

33. Addition: Two Numbers

4	7	5	1	
9	7	3	1	
4	6	0	8	2
(i)	(ii)	(iii)	(iv)	(v)

In (i) 4 must be 1. And since (ii) cannot be $\underline{9}$ (for 9 is the figure given),

.˙. it must be $\underline{8}$ and there is 1 to carry from (iii).

.˙. in (iii) 7 $\overline{\underline{\text{must}}}$ stand for 5 or more. But not 5, for the last figure in (iii) would then be 0 or 1 (both impossible). And 7 cannot stand for 8,

147

for 9 stands for 8. And 7 cannot stand for 9, for the bottom figure in (iii) would then be 8 (which we have) or 9, and we are assuming that 7 stands for 9. . ˙. (iii) must be 6

$$\frac{6}{2 \text{ or } 3}$$

The bottom figure in (v) (2) must be even, and it cannot be 0 or 2 or 6 or 8. . ˙. it must be 4.

. ˙. (v) must be $\frac{2}{4}$ or $\frac{7}{4}$.

If it were $\frac{2}{4}$, then the digits in (iv) would be 5, 7 and 9;

and there would *not* be 1 to carry from (v) to (iv). But without 1 to carry, two odd digits will add up to an *even* one, *not* to an odd one. . ˙. this is not possible.

. ˙. (v) must be $\frac{7}{4}$.

. ˙. (iv) must be 2 or 3 and 5 and 9 (the only digits left). 9 + (2 or 3) + 1 (carried) = 12 or 13 (not 15); 9 + 5 + 1 (carried) = 15 (not 12 or 13).

But if 9 is below the line, then there cannot be anything to carry and we can have 5 + 3 + 1 (carried) = 9. But since 5 + 3 are the figures given, it must be 3 + 5.

Complete Solution

$$\frac{\begin{array}{r} 1637 \\ 8657 \end{array}}{10294}$$

34. Only One Out

```
        7 1        (i)
   45 ) 3 3 4 3    (ii)
      1 2 5        (iii)
      ―――――
        8 3        (iv)
        8 1        (v)
      ―――――
```

The second figure of (i) cannot be 0 and must therefore be 2. And the second figures of (iv) and (v) must be 2. ∴ (iv) and (v) are the divisor times 2, ∴ the divisor must end in 1 or 6. But not 1, ∴ 6.

The first figure in the divisor cannot be 5, for (iv) and (v) would then have three figures (5_ × 2 = 1 _ _).
∴ the divisor is 36.

The first figure in (1) is 6 or 8. *Suppose 8.* Then 36 × 8 = 288. But the figures in (iii) are not all 1 out.
∴ (i) is 62, and (iii) is 216.

Complete Solution

```
        6 2
   36 ) 2 2 3 2
      2 1 6
      ―――――
        7 2
        7 2
      ―――――
```

35. Nearly Right

We must first look for the figure that is incorrect: if A's played were changed from 3, then one of the other figures would also have to be changed. And the same thing applies to A's lost, drawn and points.

And this is also true of the played, won, lost, drawn and points of B, D and E.

But it is *not* true of C; ∴ the mistake must be in C's won or lost, or in one of the goals for or goals against. And all the other figures are correct.

The total of the matches played must be even, for each match appears twice.. ˙. C must have played 2 or 4. The total of points must be *even*, for 2 points are given for each match, and the total is equal to that of matches played.. ˙. C's points must be 3 (if C played 2) or 5 (if C played 4).

Suppose C played 2 and got 3 points. We know that E drew none (figure given) and D cannot have drawn any, for they got no points.

˙. A's 2 drawn matches must have been v. B and v. C.. ˙. C's other match must have been v. B, who played them all, and C would have lost this match, for B won *all* their matches except the one they drew v. A.. ˙. C could not have got 3 points.. ˙. C must have played 4 and got 5 points.

Consider D. They got no points, ˙. they lost both their matches and they must have been v. B and v. C, who played every one.. ˙. D did not play A or E.. ˙. E played A, B and C. We know that A lost v. E, for A got their 2 points by 2 draws v. B and v. C.. ˙. E lost v. B and v. C (for E only got 2 points). And since B and C both played 4 matches, they played each other, and B won the match, for they got 7 points.

So we have:

	A	B	C	D	E	GOALS FOR	GOALS AGAINST
A		Dr	Dr		L		
B	Dr		W	W	W	5	
C	Dr	L		W	W		2
D		L	L			3	5
E	W	L	L			3	2

And the mistake is in C's wins. It should be 2, not 1.. ˙. every other figure is correct.

E lost 2 (v. B and v. C) and had 2 goals against.. ˙. E v. B was 0–1 and E v. C was 0–1.. ˙. E v. A was 3–0.

C had 2 goals against. None by E, but at least 1 by B who beat them. .˙. C v. D was 2–1 or 1–0 (D had 3 goals for and 5 against and therefore lost both their matches by a single goal). If D v. C was 0–1, then D v. B would be 3–4. But B v. D cannot be 4–3, for B scored 1 v. E, and at least 1 v. C (they won this match), ∴ not more than 3 v. D as B scored 5 goals for. ∴ C v. D must have been 2–1.

∴ D v. B was 2–3.

∴ from B's goals for, B v. C was 1–0 and B v. A was 0–0. And from C's goals against C v. A was 0–0.

Complete Solution

The figure that was wrong was C's WON; it should be 2, not 1.

A v. B : 0–0
A v. C : 0–0
A v. E : 0–3
B v. C: 1–0
B v. D: 3–2
B v. E : 1–0
C v. D: 2–1
C v. E : 1–0

36. One Wrong Again

Let us first try to find out more about the figure that is wrong. If the figures given for matches played were correct, then B's played would have to be 2 (to make total even). But if B played 2 they could not have drawn 2 and got 4 points (and had 5 goals for and 3 against).

∴ the mistake must be *either* in matches played *or* in details for B. ∴ all else is correct.

A got 5 points, ∴ they must have played 3, won 2 and drawn 1. ∴ A's played is correct. ∴ A played C. And since A only had 3 goals against and C had 6 goals for, ∴ C must have played more than 1 game. ∴ C played 2 or 3.

Suppose B's drawn (2) is wrong. Then all else is correct. ∴ B's points (4) is correct, and if B did not draw 2 they can only have got 4 points by winning 2. And since they lost none, they can only have played 2. And since they won both their matches they cannot have played A. But A played them all. . ˙. B's drawn (2) must be correct. And since D got no points, they drew none, . ˙. B v. A and B v. C were both drawn. And A won their other 2 matches v. C and v. D.

If B's points were wrong and all else is correct, then *either* B played only 2 and goals for and against are wrong, *or* B must have lost 1 (not 0 – figure given). . ˙. B's points must be correct and B v. D was a win for B.

We can now see that *either* C's played *or* D's played must be the incorrect figure, for the sum of the matches played should be even.

Consider C v. D. If this match was played, then C would have won it, for D got *no* points. But C did not win a match. . ˙. C did not play D. . ˙. D's played is the incorrect figure, it should be 2, not 3.

We now know who played whom and the result of each match.

Consider C v. B (a draw). B only had 3 goals against, . ˙. the score in this match cannot have been more than 3–3.

Suppose C v. B was 2–2. Then the score in C's other match (v. A) would have been 4–6. But this is not possible, for A only had 3 goals against. And if the score in C v. B was *less* than 2–2, the situation would be worse.

. ˙. C v. B can only be 3–3, . ˙. C v. A was 3–5.

And since A only had 3 goals against, . ˙. A v. B was 0–0, and A v. D was ?–0. Since B had 5 goals for and 3 against, . ˙. B v. D was 2–0. And since D had 8 goals against, D v. A was 0–6.

Complete Solution

A v. B : 0–0
A v. C: 5–3
A v. D: 6–0
B v. C: 3–3
B v. D: 2–0

37. Football: Five Teams, Two Figures Wrong

Consider D. If they won 2 and drew 1, they would have got 5 points – not 6. . ˙. one of D's figures must be wrong.

Consider A. If they played 3 and drew 2 and lost 0, then they must have won 1 and their goals for could not be equal to their goals against. . ˙. there is a mistake in D's figures and in A's figures. . ˙. all other figures are correct.

B lost 3 and had 3 goals for and 5 against. . ˙. since (5 − 3) is only 2, B must also have won 1. . ˙. B and E both played all the other four teams.

Since C only played 2, they only played B and E. D did not play C, so they only played 2 or 3. If they only played 2 then at least *two* of D's figures would be wrong (their points *and either* their wins *or* their draws). . ˙. D must also have played A, and we know that A's played (3) is correct. We now know who played whom.

Since D played 3, their points could not be 6 unless their wins and draws were both wrong. . ˙. D's points must be D's incorrect figure and it should be 5, not 6.

Suppose A's draws is A's incorrect figure. Then A would have to have won 3 (the total of drawn matches must be even, A's lost figure cannot also be changed, and B cannot have drawn 1; see their goals for and against). But if A won 3, then wins would be 8, and lost would be 6. (Remember that C played 2, drew 0, and got 2 points. . ˙. they won 1 and lost 1.) And similarly, if A's losses were changed, the total of wins and losses would not be the same. . ˙. the mistake for A must be in A's goals for or against.

Since A drew 2, . ˙. A v. D was a draw, and A v. E was a draw. And A v. B was a win for A.

D won their other 2 matches, v. B and v. E.

C won 1 and lost 1, and total score was 6–3. . ˙. C won a match by at least 4 goals. But this would not have been v. B, who only had 5 goals against, at least 3 of which must have been in the 3 matches which they lost. . ˙. C beat E and lost to B. . ˙. B v. E was won by E.

We now know the result of each match.

Consider E. They had 5 goals against. C won their match v. E by at least 4 goals, and E was also beaten by D. . ˙. E v. C was 0–4, and E v. D was 0–1. . ˙. E v. A was 0–0, and E v. B was ?–0. Since C v. E was 4–0, . ˙. C v. B was 1–2.

Consider B. Since B v. C was 2–1, and B's goals altogether were 3 for and 5 against, . ˙. each of B's 3 lost matches were lost by 1 goal. . ˙. B v. E was 0–1. B v. D must be 1–2 or 0–1. *Suppose it were 0–1.* Then D v. A would then be 5–5 (D had 5 goals against). But this is not possible, for A's goals for and against would then both be wrong. . ˙. B v. D must be 1–2, and B v. A is 0–1. Since D v. B is 2–1, . ˙. D v. A is 4–4.

Complete Solution

The wrong figures are:

    (i) A's goals for – they should be 5, not 4.
(ii) D's points – they should be 5, not 6.

A v. B : 1–0
A v. D: 4–4
A v. E : 0–0
B v. C: 2–1
B v. D: 1–2
B v. E : 0–1
C v. E : 4–0
D v. E : 1–0

38. Uncle Bungle and the Vertical Tear

B must have got at least 1 point (x goals for and x goals against). If B got 2 points, then they would either have drawn 2 matches or would have won 1 and lost 1 (they could not have got more than 2 points).

Since only *some* of the matches had been played, only 4 points can have been obtained for 2 matches (not 1 match, for all three teams had goals for or against or points). If B got 2 points, then A and C would have got 2 points between them. But this could only have been 0 and 2, or 1 and 1; and in each case two teams would have got the same number of points. But they are different, ∴ B's points (p) must be 1.

And A's and C's points (m and x) must be 0 and 3 or 3 and 0.

Suppose that $m = 3$ and $x = 0$. Then C would have got no points and had no goals scored against. But this is not possible, for g would then be greater than 0 and C would therefore have got some points.

∴ m must be 0 and x must be 3. ∴ C with 3 points must have played 2 matches, and these were the only matches played. A v. C was 0–1 and B v. C was 3–3, ∴ g (C's goals for) was 4.

Complete Solution

    A v. C: 0–1
    B v. C: 3–3

39. Uncle Bungle and the Horizontal Tear

It will only be possible 'to discover the score in each match' if the figures for C's matches played, won, lost, drawn, etc., are the same as D's. For otherwise there could be no reason why the figures for C and D should not be interchanged; but if they are the same it will not matter if they are interchanged. On this assumption, *and only on this assumption*, it will be 'possible to discover the score in each match'.

A drew 2 matches, ∴ A v. C and A v. D must both be draws, and the *scores must be the same*. The scores could not be 0–0, for A v. B would then be 7–5 (too many). Nor could the scores be 1–1, for A v. B would then be 5–3 (too many). Nor could the scores be 3–3, for A only had 5 goals against, not 6. ∴ the scores in A v. C and in A v. D were 2–2.

∴ the score in A v. B was 3–1.

∴ B scored 4 goals v. C and D (5 minus 1), and B had 2 goals scored against them by C and D (5 minus 3). ∴ the scores in B v. C and in B v. D were 2–1.

Complete Solution

    A v. B : 3–1
    A v. C: 2–2
    A v. D: 2–2
    B v. C: 2–1
    B v. D: 2–1

40. The Lie Drug

Remembering that all figures are 1 out, we can say that A, B and C's figures must be:

	PLAYED	WON	LOST	DRAWN	GOALS FOR	GOALS AGAINST	POINTS
A	3	1	1	1	5 or 7	6 or 8	3
B	2	1	0	1	1	0	3 (not 2 against, for B won 1 and drew 1)
C	2	0	1	1	2	4	1 (goals against must be greater than goals for)

D could have played 1 or 3. If they played 3, they must either have won 2, lost 0 and drawn 1, or won 0, lost 2 and drawn 1. If they won 2 and drew 1, they would have got 5 points, but we know that they only got 1 point. And if they lost 2 and drew 1, they would have had more goals against than goals for. But their goals are 3 or 5 for, and 1 or 3 against.

∴ D cannot have played 3, ∴ they played 1. And their figures are:

	PLAYED	WON	LOST	DRAWN	GOALS FOR	GOALS AGAINST	POINTS
D	1	0	0	1	3	3	1

A played 3, ∴ D's match was v. A, and the score was 3–3.

∴ the other drawn match was B v. C, and since B had no goals against, the score was 0–0. And B's match v. A must have been 1–0.

C v. B was 0–0, ∴ C v. A (C's only other match) was 2–4.

Complete Solution

A v. B : 0–1
A v. C: 4–2
A v. D: 3–3
B v. C: 0–0

41. Uncle Bungle and a Lucky 13

	—	—	—	—		(i)
				—	t	(ii)
p	x	k	y	p		(iii)
r	y	d	x			(iv)
p	p	t	t	m	p	(v)
(vi)	(vii)	(viii)	(ix)	(x)		

p, the first figure in (v) can only be 1... r in (vi) must be 9 for $1 + r = 11$, and there must be 1 to carry from (vii).

In (ix) $y + x = m$, or $m + 10$ (there cannot be anything to carry from (x)).

In (vii) and (ix) the same two numbers are added together (x and y). .·. $t - m$ must be 1 (we know that p is 1 and that r is 9, .·. t and m are neither 1 nor 9, nor can either of them be 0, for t could only be 0 (i.e. 10 with 1 carried) if m were 9.

It is not possible for k or d to be 0, for there could not then be anything to carry from (viii). .·. x, y, k, d, t and m must be 6 of the digits 2, 3, 4, 5, 6, 7 and 8.

Suppose that y and x were 7 and 8. Then m would be 5 and t would be 6. The most that k and d could then be is 3 and 4, which is not enough (there would be nothing to carry).

Suppose that y and x were 6 and 8. Then m would be 4 and t would be 5, and the most that k and d could then be is 7 and 3, which would make t in (viii) 1 instead of 5.

Suppose that y and x were 8 and 5. Then m would be 3 and t would be 4. And k and d could then be 6 and 7, which would make t in (viii) 4. But this would only be all right if (v) (114431) were a multiple of 13. But if we divide we see it is not.

157

Suppose that y and x were 7 *and* 5. Then *m* would be 2 and *t* would be 3, and *k* and *d* could then be 8 and 4. (v) would then be 113321, and this is the only other possibility. If we divide this by 13, we see that it goes exactly 8717 times. ∴ (v) can only be 113321.

We now know that *t* = 3 and *m* = 2.

(iv) is less than (iii), so the first digit in (ii) must be 1 or 2.

Suppose it were 1. Then the first digit in (i) would be 9 (for this is the first digit in (iv)). But 3 × 9 is 27, and (iii) would not then start with 1 as it should. ∴ (ii) can only be 23.

In order to get (i) we must divide 113321 by 23. And this gives us 4927. 3 × 4927 = 14781; 2 × 4927 = 9854.

Complete Solution

```
    4927
      23
   14781
    9854
  113321
```

42. No Divisor

		g	*x*	*a*	*p*	*m*	*g*	(i)
)	*m*	*m*	*d*	*p*	*b*	*p*	*g*	(ii)
	m	*g*						(iii)
		g	*d*	*p*				(iv)
		g	*m*	*a*				(v)
				b	*b*			(vi)
				b	*p*			(vii)
					p	*p*		(viii)
					p	*m*		(ix)
						m	*g*	(x)
						m	*g*	(xi)

(The reader is advised to have a figure, like the one on the right above, where the digits can be inserted as they are found.)

From (ii), (iii) and (iv) $g + g = m$, $\therefore m = 2g$.

From (viii), (ix) and (x) $m + m = p$, $\therefore p = 2m = 4g$.

From (vi), (vii) and (viii) $p + p = b$, $\therefore b = 2p = 4m = 8g$.

$\therefore g = 1, m = 2, p = 4, b = 8$. And since the last figure of (i) is 1, and (xi) is 21, $\therefore$ the divisor is 21. Since in (v) $m = 2$, $\therefore d = 3$. Since (v) is 12, $\therefore a = 6$. And x (the second figure in the dividend) is 0.

Complete Solution

The divisor is 21.

```
          106421
    21)2234841
      21
       134
       126
         88
         84
          44
          42
           21
           21
```

43. Addition: Letters for Digits, One Letter Wrong

T	L	S	L	T	T	E	M	A
T	L	S	T	B	T	T	M	K
B	B	S	H	L	M	D	S	A
(i)	(ii)	(iii)	(iv)	(v)	(vi)	(vii)	(viii)	(ix)

Look first for the incorrect letter.

Consider (iii). If this is correct, then S must stand for 0 or 9 (with 1 that has been carried from (iv)). If S = 0, then (ix) cannot be right, for it is only right if K = o. If the mistake is in (ix), then everything else is correct. ∴ all S's in (iii) are o, and there is nothing to carry to (ii).

Let us see whether this is possible. We have

$$\frac{TL}{BB}$$

And L − T must be 5, so that 2 L − 2 T is 10. ∴ L is greater than 5, ∴ B is odd, ∴ there must be something to carry from (iii). ∴ S cannot be o, for if it were there would be a mistake in (i) or in (ii) *and* one in (ix). ∴ if (iii) is correct, S would have to be 9; but in that case the S in (viii) cannot be correct, *unless* there is a mistake in (ix).

∴ the mistake must be in (iii), (viii) or (ix). ∴ there is no mistake elsewhere.

∴ (i) and (ii) must be $\dfrac{16}{33}$ or $\dfrac{27}{55}$ or $\dfrac{38}{77}$; *not* $\dfrac{49}{99}$,

for L and B would then be the same.

But from (v): $\dfrac{T}{\underline{L}}$ they can only be: $\dfrac{27}{\underline{55}}$ ∴ T = 2, L = 7, B = 5.

In (iv) H = 9, ∴ S is not 9; ∴ the mistake must be in (iii). ∴ (viii) and (ix) are correct.

∴ in (ix) K = o. In (vi) M = 4 (it cannot be 5, so there is nothing to carry from (vii)). ∴ in (viii) S = 8.

The only digits left now are 1, 3 and 6. ∴ in (vii) E = 1, D = 3 (1 + 2 = 3). ∴ in (ix) A = 6. S in (viii) is 8. ∴ *two* of the S's in (iii) must be 8;

$$\frac{8}{\underline{?}}$$
$$\underline{8}$$

is not possible, for since there is nothing to carry from (iv), ? cannot be 9; since there *is* something to carry to (ii), ? cannot be o.

∴ we can only have $\dfrac{8}{\underline{6}}$.

Complete Solution

The third S in (iii) is wrong; it should be 6 (not 8).

$$\begin{array}{c}
2\,7\,8\,7\,2\,2\,1\,4\,6 \\
2\,7\,8\,2\,5\,2\,2\,4\,0 \\
\hline
5\,5\,6\,9\,7\,4\,3\,8\,6
\end{array}$$

44. Uncle Bungle Gets the Last Line Wrong

(i)	(ii)	(iii)	(iv)	(v)	(vi)	(vii)	(viii)	
E	X	M	R	E	E	K		
E	H	K	R	E	K	K		
− K	−	− H	−	− X	−	− E		(ix)

There are six different letters above (E, H, K, M, R and X), and since all 10 digits are included the blanks must all be different from each other and different from the digits for which the letters stand. The most that can be carried when two digits are added together is 1.
∴ (i) must be 1.

From (viii) E is even (K + K); and from (i) and (ii) must be 5 or more.
∴ E is 6 or 8. If E were 8, then K in (viii) would be 4 or 9. And if E were 8 in (ii), then K would be 6 or 7.

∴ E cannot be 8, and must be 6. ∴ K in (ii) and (viii) must be 3. And there is 1 to carry from (iii) to (ii).

Since E is 6 and K is 3, ∴ the seventh figure in (ix) is 9. And in (vi) X is 2 (6 + 6 = 12), and there is 1 to carry to (v). Since there is 1 to carry from (iii) to (ii), ∴ H must be at least 7. But if H is 7 or more there cannot be anything to carry from (iv). ∴ H can only be 8 (not 9, for the seventh digit of (ix) is 9). ∴ the third digit of (ix) is 0. M in (iv) must be 4 or 5.

Suppose M were 5. Then R in (v) would be 4 or 7 (the only digits left). But not 4, for the fifth digit of (ix) would then be 9, but this is not possible for the seventh digit of (ix) is 9. ∴ A and R could not be 7, for we are assuming that there is nothing to carry from (v).

∴ M cannot be 5. ∴ M must be 4. And there *is* 1 to carry from (v).
∴ R = 7, and the fifth digit of (ix) is 5.

Complete Solution

```
  6 2 4 7 6 6 3
  6 8 3 7 6 3 3
1 3 0 8 5 2 9 6
```

45. Bungled Again

| | H | B | B | D | B | M | D | B |
	H	B	B	B	P	A	D	B
M	B	H	E	G	T	X	B	H
(i)	(ii)	(iii)	(iv)	(v)	(vi)	(vii)	(viii)	(ix)

We must first try to find more about the letter that is wrong, remembering that it is in the last line across.

H in (ii) must be 5 or more to produce the first of the (ix) letters in the third row across. If H were 5, then the mistake would be in (ix) (B + B must be even). .˙. everything else would be right.

.˙. B in (ii) would have to be 0 or 1; but not 0, for then in (iii) H would be 0 or 1, and we are assuming that H is 5. And if B in (ii) were 1, then H would be 2 or 3. .˙. H cannot be 5.

Suppose H were 6 in (ii). Then B in (ii) would be 2 (not 3, for if B in (iii) were 2 or 3 there would not be anything to carry from (iii) to (ii)). .˙. H in (iii) would be 4 (not 5, see (iv)). And this could be one mistake. But if H were 6 in (ix), B would be 3 or 8. But B in (ii) would be 2. .˙. there would be at least two mistakes, and H cannot be 6.

If H were 7, there would be a mistake in (ix); and B would be 4 or 5 in (ii), but B would have to be 3 or 8 in (iii). .˙. H cannot be 7.

If H were 9, there would be a mistake in (ix); and B would be 8 in (ii) (not 5 for H is 9), but there would be 1 to carry from (iii), making B 9.

.˙. H can only be 8.

.˙. in (ix) B must be 4 or 9. And in (ii) B would have to be 6 or 7. .˙. the mistake is in (ii) or (ix) and everything else is correct.

.˙. in (i) M = 1. B in (iii) must be 4 or 9, and there cannot be one to carry from (iv) for 8 is even. .˙. E in (iv) can only be 9, and there is 1 to carry from (v). Since E is 9, .˙. B must be 4. .˙. D in (viii) is 2 or 7. But not 2, for in (v) if D were 2 there would not be 1 to carry to (iv). .˙. D = 7. In (v) G must be 1 or 2, but not 1, for M is 1. .˙. G = 2. In (vii) M is 1, and since A cannot be more than 6, .˙. there cannot be 1 to carry to (vi). .˙. in (vi) P must be 6, and T must be 0 (we know that there is 1 to carry from (vi) to (v)). .˙. in (vii) A must be 3, and X must be 5. .˙. the mistake is in (ii). B should be P (6).

Complete Solution

The B in (ii) is wrong; it should be P (6).

```
  84474174
  84446374
 168920548
```

46. H is 3

	Y	X	P	(i)
			H	
P	M	Y	X	(ii)

The most that can be carried when a number is multiplied by 3 is 2
$(3 \times 9 = 27)$. ∴ P in (ii) must be 1 or 2.

If P were 1, then X in (ii) would be 3 $(3 \times 1 = 3)$. But H and X are different.

∴ P can only be 2. ∴ X in (ii) is 6. ∴ Y in (ii) is 8 $(3 \times 6 = 18)$.
∴ M in (ii) is 5 $((3 \times 8) + 1 = 25)$.

Complete Solution

```
  862
    3
 2586
```

47. Rows of Eight

D H G D N G N C	(i)
C H C D N D G D	(ii)
E P B N B B G E	(iii)

From the first digits of (i), (ii) and (iii) there is not 1 to carry when D and C are added. .·. there is not 1 to carry from *last* digits of (i) and (ii). .·. D + C = E.

Consider the seventh column.
$$\begin{array}{r} N \\ G \\ \hline G \end{array}$$
N must be 0.

Consider the fifth column.
$$\begin{array}{r} N \quad\quad o \\ N, \text{ i.e. } o \\ \hline B \quad\quad B \end{array}$$
since B is not 0, .·. it must be 1.

Consider the fourth column.
$$\begin{array}{r} D \\ D \\ \hline N \end{array}$$
since N = 0, .·. D = 5.

And from the sixth column G = 6; and from the third column C = 4 .·. from the last column E = 9. And in the second column (considering which digits there are left) H must be 3, and P must be 7.

Complete Solution

```
53650604
43450565
97101169
```

48. Rows of Nine

M	M	W	X	T	F	G	G	G	(i)
M	M	E	X	W	T	F	G	G	(ii)
M	M	Y	F	M	M	F	G	G	(iii)
F	T	T	Y	M	C	V	F	M	(iv)

Consider the first figure of (i), (ii), (iii) and (iv). M must be 1, 2 or 3 (otherwise there would be another figure in (iv)). If M = 3, then G would be 1, and F would also be 3. .·. M is not 3.

If M = 2, then the first figure of (iv) (F) would be 6 (there could not be one to carry from the second column (M + M + M)). And from the last digits down G would be 4 (3 times 4 = 12). ∴ the last digit but one of (iv) would be 3 (3 times 4 + 1).

But according to our assumption F = 6, not 3.

∴ our assumption is wrong. ∴ M is not 2. ∴ M = 1, ∴ G = 7 (3 times 7 = 21), and F = 3.

From the seventh column (7 + 3 + 3 + 2 (that has been carried)) is followed by V. ∴ V = 5 (and 1 has been carried).

∴ from the second column T must be 4 (since V = 5). From the sixth line down 3 + 4 + 1 + 1 = C, ∴ C = 9; and from the fifth column W = 6.

Consider the third column. 6 + E + Y + (perhaps) 1 = 14, and 0, 2 and 8 are only digits left.

It is easy to see that E and Y must be 0 and 8 (though we cannot yet say which is which).

∴ X = 2, and from the fourth column Y = 2 + 2 + 3 + 1 = 8. ∴ E = 0.

Complete Solution

$$
\begin{array}{r}
116243777 \\
110264377 \\
118311377 \\
\hline
344819531
\end{array}
$$

49. Multiplication of Seven Digits

$$
\begin{array}{r}
\text{A T V T S A V} \qquad \text{(i)} \\
\text{A} \\
\hline
\text{V V J E C K S A} \qquad \text{(ii)}
\end{array}
$$

Since A is 5 or less, ∴ the first figure of (ii), (V) cannot be more than 2 ((5 × 5) + 4 = 29). But it is not possible for V to be 2, for 5 × 2 = 10, 4 × 2 = 8, 3 × 2 = 6, 2 × 2 = 4, and in no case is the last digit of (ii) the same as the multiplier (A cannot be 1).

∴ V = 1 (not 0, for the last digit of (ii) would then be 0). ∴ the first two digits of (ii) are 11, and A must be 3 (3 × 3 = 9, and there must be 2 to carry to make 11).

S = 9 (last digits but one in (ii), 3 × 3 = 9). ∴ K = 7 (3 × 9 = 27).

Consider T (the second digit of (i)). Since there must be 2 to carry after multiplying by 3, ∴ T must be 7, 8 or 9. But not 7 or 9 (they are K and S), ∴ T = 8.

∴ we now know all of (i). Multiply by 3, and we get C = 6, E = 5 and J = 4.

Complete Solution

```
3 8 1 8 9 3 1
            3
1 1 4 5 6 7 9 3
```

50. Two Rows

```
A X S X V A E      (i)
X S S A V A E      (ii)
W H D T X E E      (iii)
```

Consider the last column − E + E = E. E must be 0.

Consider the last column but one −

$$\frac{\begin{array}{c} A \\ A \end{array}}{E}$$

A must be 5, and note there is 1 to carry.

Consider

$$\frac{\begin{array}{c} V \\ V \end{array}}{X}$$

Since 1 has been carried, ∴ X is odd. But X is not 1, for V would then be 5, and A is 5. And X cannot be more than 4, for from the first digits of (i), (ii) and (iii) 5 + X = W, and W must be less than 10.

∴ X = 3, and V = 1 or 6.

We now have:

```
5 3 − 3 − 5 0
3 − − 5 − 5 0
− − − − 3 0 0
```

Suppose V = *1*. Then we have 3 + 5 = T; .˙. T = 8 (not 1 to carry, for V = 1)..˙. in the first line down 5 + 3 = W, .˙. W must be 9 (not 8, for T = 8).

$$\begin{array}{c} X \\ \text{Consider} \quad S \\ \underline{H} \end{array}$$

X = 3, and S cannot be more than 7. If S = 7, H = 0 or 1. But E = 0 and V = 1.

.˙. our assumption is incorrect. .˙. V = 6, .˙. T = 9, and W = 8.

$$\begin{array}{c} S \\ \text{Consider} \quad S \\ \underline{D} \end{array}$$

D must be even, but not 6 or 8. .˙. 2 or 4. *Suppose* 4, then S = 2, and X + S = H becomes 3 + 2 = 5. But A is 5, not H.

.˙. our assumption is incorrect, and D = 2. .˙. S = 1, and H = 4.

Complete Solution

$$\begin{array}{r} 5\,3\,1\,3\,6\,5\,0 \\ 3\,1\,1\,5\,6\,5\,0 \\ \hline 8\,4\,2\,9\,3\,0\,0 \end{array}$$

51. Five Digits Divided by Two Digits

		m a h q	(i)		----
a g	)	*d b a r p*	(ii)	--)-----
		m q	(iii)	--
		g a	(iv)	--
		a g	(v)	--
		j r	(vi)	--
		q m	(vii)	--
		j p	(viii)	--
		j p	(ix)	--

(The reader is advised to have a figure, like the one on the right above, where the digits can be inserted as they are found.)

(v) is the divisor, .˙. a = 1 (the second figure in (i)). From (iv) and (v), g = 2.

From (iv), (v) and (vi), $j = 9$, ∴ from (vii) $q = 8$. ∴ (ix) is 12 × 8, i.e. 96, ∴ $p = 6$. From (ii), (iii), and (iv) $b = 0$.

In (vii) 8 – must be 12 × 7, ∴ $h = 7$ and $m = 4$ ∴ in (ii) $d = 5$. From (vi) (vii) and (viii) $r = 3$.

Complete Solution

```
         4 1 7 8
12 ) 5 0 1 3 6
     4 8
     ‾‾‾
       2 1
       1 2
       ‾‾‾
         9 3
         8 4
         ‾‾‾
           9 6
           9 6
           ‾‾‾
```

52. Addition: Two Numbers

	X	D
H	H	D
X	D	H
(i)	(ii)	(iii)

From (iii) H is even (D + D).

And from (i) X = H + 1 (not more than 1 can be carried from (ii)).

Since 1 is carried from (ii), ∴ X + H + (perhaps) 1 from (iii) is 10 or more.

∴ H is not 2. If H were 4, and X were 5, 1 would have to be carried from (iii), but this is not possible, for D would then be 0.

Suppose H were 8, and X were 9. Then D in (iii) would be 4 or 9. But in (ii) X + H + (perhaps) 1, would be 17 or 18. ∴ H is not 8.

∴ H must be 6, and X = 7 and D = 3.

Complete Solution

```
  7 3
6 6 3
7 3 6
```

53. Multiplication of Five Digits

```
R E M H B      (i)
        M      (ii)
T M G B R      (iii)
```

M in (ii) cannot be 0 or 1. *Suppose it was 3.* Then B in (i) is at least 2.
.˙. R in (iii) would be 6. .˙. R (the first figure in (i)) would be 6, and
there would be another figure in (iii). And if M in (ii) were more than
3, the situation would be worse. .˙. M in (ii) must be 2.
   .˙. R in (iii) must be even. It cannot be 0, for R in (i) cannot be 0.
.˙. R must be at least 4. And it cannot be more than 4, for in that
case there would be another figure in (iii) (for R is the first figure in
(i)). .˙. R = 4. And in (i), since B cannot be 2, it must be 7. .˙. B = 7.
.˙. B in (iii) is 7.
   .˙. H in (i) is 8 ($2 \times 8 = 16$, and 1 carried makes 17).
   .˙. G in (iii) is 5 ($2 \times 2 = 4$, and 1 carried makes 5).
   We know that T in (iii) must be 8 or 9 ($2 \times 4 = 8$). But since H is 8, T
must be 9 and there must be 1 to carry.
   .˙. E in (i) is 6 ($2 \times 6 = 12$).

Complete Solution

```
4 6 2 8 7
        2
9 2 5 7 4
```

54. More Addition-by-Letters

```
A A R X B A A
B A G X A A A
B P A X B A A
D H A R H R B
```

Consider the last column. (A + A + A) is followed by B.

Consider the last column but one. (A + A + A) is followed by R. ∴ there must have been 1 or 2 to carry from the last column, ∴ A is greater than 3.

Consider the first column. A + B + B + (perhaps 1 or 2) = D. ∴ A cannot be 9. If A was 8, then B would be 4 – impossible. If A was 7, then B would be 1, but there would be at least 1 to carry from the second column (7 + 7 + ?), so that the first column would be at least (7 + 1 + 1 + 1).

If A was 6, B would be 8 – impossible. If A was 5, B would be 5. ∴ A = 4, ∴ B = 2, ∴ from the last column but one, R = 3. ∴ from the last column but two, H = 9, ∴ (X + X + X) = 3, ∴ X = 1.

The first column = (4 + 2 + 2). We know that D is not 9, ∴ D = 8.

In the third column 3 + G + 4 = 14, ∴ G = 7, ∴ P = 0.

Complete Solution

```
4 4 3 1 2 4 4
2 4 7 1 4 4 4
2 0 4 1 2 4 4
8 9 4 3 9 3 2
```

55. Addition: Four Numbers

```
        E       (i)
      K X       (ii)
    Y K E       (iii)
    Y Y K X     (iv)
  E E P D K     (v)
```

The first figure of (v) must be 1 (the most that can be carried forward from the second column). .˙. E = 1.

.˙. (v) starts 11; and Y must be 9, and 2 must have been carried from the third column (Y + Y).

.˙. P must be 0 (9 + 9 + 2 = 20); and even if it were possible for 3 to be carried from the fourth column, P cannot be 1, for E is.

.˙. there is 2 to carry from the fourth column, (K + K + K), and from the last column we know that K is even (1 + X + 1 + X). .˙. K = 8 (not 6, for though 6 + 6 + 6 + 2 = 20, and 6 + 6 + 6 + 3 = 21, 0 and 1 are P and E).

The last column adds up to 8, 18 or 28. But *not* 28, for since E = 1, X would be 13, which is impossible. And *not* 18, for X would then be 8 (1 + 8 + 1 + 8 = 18), and K = 8. .˙. the last column adds up to 8, X = 3, and D = 4.

Complete Solution

```
      1
     8 3
    9 8 1
    9 9 8 3
  1 1 0 4 8
```

56. A Multiplication

B	X	Y	D	X	B	T		(i)
						T		(ii)
B	M	X	R	Y	P	D	T	(iii)

Consider the last letters in (i) and (iii) and T in (ii): T × T can only produce –T, if T is equal to o, 1, 5 or 6. o and 1 are clearly not possible here. . `. T must be 5 or 6.

Suppose T were 5. If B were 9, (iii) would start with 45 or rather more. If B were 6, (iii) would start with 30 or rather more. We can see from these examples that the smaller B is, the more likely it is that we shall have B at the beginning of (iii).

Suppose B were 1 (and T were still 5). Then (i) could start 198 and (iii) would then start 990, but it would then only have 7 figures instead of 8. And we can see that it is not possible for T to be 5. . `. T can only be 6.

If B were 2 and X were 9, (iii) would start 174, but it should start with 2. . `. B can only be 1.

Since T is 6, there must be 3 to carry from the last column (6 × 6 = 36). And since the last digit but one in (i) is B (1), D in (iii) is (6 + 3), i.e. 9. If (i) started 187, (iii) would start 1122. But M cannot be 1, for B is 1. . `. M can only be o.

. `. (iii) starts 10, and B is 1, . `. 4 has been carried (6 × 1 = 6, and 4 makes 10).

. `. X (the second letter in (i)) must be 8 or 7 (not 9, for D is 9; not 6, for T is 6; and it is easy to see that it cannot be less than 6).

If X (the fifth letter in (i)) were 8, then P (the sixth letter in (iii)) would be 8 (6 × 8 = 48, and we know that there is nothing to carry from the next column). But P and X cannot both be 8.

. `. X can only be 7. And since 6 × 7 = 42, P = 2. 6 × 9 = 54, and 4 makes 58. . `. Y = 8. And 6 × 8 = 48, and 5 makes 53. . `. R = 3. 6 × 7 = 42, and there is 5 to carry, . `. X = 7, which is as it should be.

Complete Solution

1	7	8	9	7	1	6
						6

10738296

57. Division: Some Missing Figures

```
        --           (i)
     --)---7          (ii)
       -5             (iii)
        --            (iv)
        ==            (v)
```

If – 5 (iii) is subtracted from – – –, the result is one figure.
∴ the first figure in (ii) is 1, and (iii) must be 95. 95 = 19 × 5, ∴
the divisor must be 19 and the first figure in (i) is 5.

Add up from the bottom and we get:

Complete Solution

```
        5 3
    19)1007
        9 5
        5 7
        5 7
```

58. A Division Sum

```
       2--           (i)
    --)----7          (ii)
       --             (iii)
      4--             (iv)
      ---             (v)
       --             (vi)
       ==             (vii)
```

Since (vi) and (vii) end in 7 (last figure in (ii)), ∴ the divisor is odd.
Since (iii) (two figures) is the divisor × 2, ∴ it is even; and since it

comes under the first three figures of (ii) . ˙ . the divisor × 3 or more has three figures.

. ˙ . (vi) and (vii) must be the divisor × 1. . ˙ . the divisor ends in 7. If the divisor was 37 it would divide into the first two figures of (iv) and the third figure would not be needed.

And since the divisor × 2 has only two figures, . ˙ . it must be 47.

. ˙ . (iii) is 94.

(v) cannot be 47 × 8, i.e. 376, for (vi) would then have three figures. . ˙ . (v) must be 47 × 9, i.e. 423.

Add up from the bottom and we get:

Complete Solution

$$
\begin{array}{r}
2\,9\,1 \\
47\,\overline{)\,1\,3\,6\,7\,7} \\
9\,4 \\
\hline
4\,2\,7 \\
4\,2\,3 \\
\hline
4\,7 \\
4\,7 \\
\hline
\end{array}
$$

59. Long Division

$$
\begin{array}{r}
-\;-\;-\qquad\text{(i)} \\
-\;m\,\overline{)\;-\;-\;-\;-}\qquad\text{(ii)} \\
-\;p\qquad\text{(iii)} \\
\hline
p\;\;m\;-\qquad\text{(iv)} \\
-\;-\;-\qquad\text{(v)} \\
\hline
x\;-\;x\qquad\text{(vi)} \\
p\;-\;-\qquad\text{(vii)} \\
\hline
p\;-\qquad\text{(viii)}
\end{array}
$$

From the divisor and (iii), the divisor must be less than 50 (for since p and m are different, (iii) must be the divisor ×, at least, 2). . ˙ . p in (viii) is less than 5.

Suppose that the divisor starts with 1. Then (vii) could not be more than 19×9, i.e. 171. And since x in (vi) would then be 2, (viii) would be at least 31 (202 − 171), which would be more than the divisor. ∴ the divisor cannot start with 1.

Suppose that p were 1. Then the first figure in (i) would be odd. But not 1, for (iii) is not the divisor. If the first figure in (i) were 3, then (iii) would be a multiple of 3. But not 51 (17 × 3), for the divisor does not start with 1. ∴ (iii) could only be 81 (27 × 3), and if the divisor were 27, (v) could only be 162 (27 × 6). But x in (vi) could not then be 2. ∴ p cannot be 1.

If the divisor started with 4, then (iii) would be at least 8−. And since p is at least 2, this is not possible, for the first figure of (ii) would then be at least 2 + 8. ∴ the divisor must start with 2 or 3. If p were 3, then x in (vi) would be 4. But the divisor cannot be more than 39, and 39 × 9 = 351. But it would not then be possible for x in (vi) to be 4. ∴ p cannot be 3 (and clearly it cannot be more). ∴ $p = 2$, ∴ $x = 3$.

Suppose that the divisor were 29. Then (vii) could not be more than 261 (29 × 9). But (vi) could not then start with 3. ∴ the first figure in the divisor must be 3. ∴ the first figure in (i) must be 2, for if it were 3, (iii) would start with 9, which is not possible.

∴ the divisor must be 36 or 31 (36 × 2 = 72; 31 × 2 = 62). If 36, then (iv) would be 26−, and (v) would be 252 (36 × 7). But it would not then be possible for the first digit of (vi) to be 3.

∴ the divisor must be 31 and (iii) is 62. (vii) must be 279 (31 × 9), and the last digit in (i) is 9, and since x is 3, (viii) is 24, (iv) is 21−, ∴ (v) must be 186 (31 × 6).

Add up from the bottom and we get:

Complete Solution

```
       2 6 9
3 1 ) 8 3 6 3
      6 2
      2 1 6
      1 8 6
      3 0 3
      2 7 9
        2 4
```

60. Long Multiplication

```
  – – – –        (i)
       x  g      (ii)
  y  b  h  m  g  (iii)
  g  y  x  g     (iv)
x  d  x  d  p  g (v)
```

x, the first digit in (v), can only be 1.. ˙. x in (ii) is 1.. ˙. (iv) is (i) × 1.
. ˙. the last letter in (i) is g, and in (iii) we have $g \times g = g$ (or – g).

g cannot be 1, for x is 1, . ˙. g must be 5 or 6 (5 × 5 = 25; or 6 × 6 = 36).

Suppose that g were 5. The last two digits in (i) would be 15, and the last two digits in (iii) would be 75, and there would then be nothing to carry. y, the second letter in (iv) (and therefore in (i)), would then be 0 or 5. But not 5, for we are assuming that g is 5.. ˙. y would have to be 0. But it is not possible for y, at the beginning of (iii), to be 0.. ˙. our assumption is wrong, and g must be 6.

. ˙. the last two digits in (i) are 16, and the last two digits in (iii) are 96.. ˙. $m = 9$, . ˙. $p = 5$.

The most that y (the first figure in (iii)) can be is 4, for 69 × 6 = 414. But this is not possible, for y in (iii) and in (iv) must be the same.. ˙. y must be less than 4.. ˙. d (the second figure in (v)) must be 0 (3 + 6 + 1 = 10). (And we can see from this that y can only be 3.) . ˙. (iii) is 37896.   ˙ ˙

Complete Solution

```
    6 3 1 6
        1 6
    3 7 8 9 6
    6 3 1 6
  1 0 1 0 5 6
```

61. Division: Some Letters for Digits, Some Missing

```
              y  –  –           (i)
    – y )   –  –  –  –          (ii)
        –  b                    (iii)
        –  –  –                 (iv)
      d  –  –                    (v)
      d  –  –                   (vi)
      e  –  –                   (vii)
        d  e                    (viii)
```

Since the divisor ends in y and (iii) ends in b, (iii) is not the divisor.
∴ the divisor is less than 50. e in (vii) must be at least 1, ∴ d in (vi) must be at least 2. And the first figure in (iv) must be at least 2. ∴ the first figure in (iii) cannot be more than 7 $(2 + 7 = 9)$. ∴ the divisor cannot be more than 39. If the divisor were 19, the most that (v) could be would be 171. But d in (v) is at least 2.

∴ the first figure in the divisor must be 2 or 3. (iii) is not the divisor, ∴ y is not 1. If y were 2, then e and d would be 3 and 4 (from (vi) and (vii) $d - e = 1$).

But that is not possible, for $39 \times 9 = 351$, so that d could not be 4. ∴ y is not 2. If y were 4, then the first figure in (iii) would be at least 8. But this is not possible, for the first figure in (ii) would then have to be 10 or more.

∴ y must be 3. ∴ $e = 1$, and $d = 2$. And in (iii) $b = 9$ (3×3), ∴ the divisor must be 23 (33 would be too large). ∴ (iii) is 69. $23 \times 9 = 207$, ∴ 207 is the only possibility for (v). And since (vii) is the largest multiple of 23 with 1 as the first figure, it must be 23×8 (184). Add up from the bottom and we get:

Complete Solution

```
         3 9 8
   2 3 ) 9 1 7 5
         6 9
         2 2 7
         2 0 7
         2 0 5
         1 8 4
           2 1
```

62. Alphabetical Chairs

E cannot be next to D (next to D in alphabet); and E is not on A's right (E's sister's husband is); ∴ E is on A's left.

C cannot be next to D, ∴ C must be on A's right.

B cannot be next to C, ∴ by elimination F must be, and B is on D's right.

Complete Solution

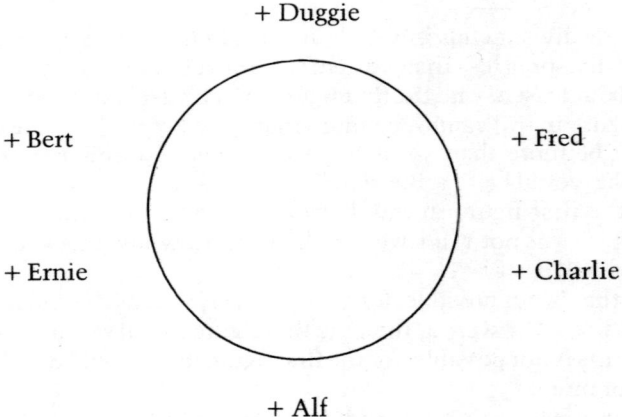

+ Duggie

+ Bert

+ Fred

+ Ernie

+ Charlie

+ Alf

63. Football in Verse

Let us first set out what we know about matches played, won, lost, etc. E played 4, and won 0; A, B and D had 3, 4 and 3 goals against. C won 1, and had no goals against and no draws; and had 4 goals for. A's matches; B's wins and B's goals for were all the same *even numbers*. Not 4, for if B won 4 and had 4 goals for, they could not have had 4 goals against. ∴ A's matches, B's wins and B's goals for were all 2.

A scored 3 goals, D had 1 draw. ∴ we have:

	PLAYED	WON	LOST	DRAWN	GOALS FOR	GOALS AGAINST
A	2				3	3
B		2			2	4
C		1		0	4	0
D				1		3
E	4	0				

E played 4 matches, ∴ E played all the others.

Consider C. They drew none, and since they had no goals against, ∴ they lost none. ∴ they only played 1, and the score was 4–0. And it must have been v. E, ∴ C v. E was 4–0.

Consider B. They won 2, and since they only scored 2 goals, ∴ the score in each of these matches was 1–0. We know that B did not play C, ∴ B can only have played 3 matches. ∴ the score in their third match was 0–4. This cannot have been v. A, for A only scored 3 goals. ∴ B v. A was 1–0.

E did not win a match, ∴ E v. B must have been 0–1 (and not 4–0); ∴ B v. D was 0–4.

Consider A. They only played 2 matches. One of them was v. B (0–1), and the other one was v. E, and the score must have been 3–2. (And A did not play D.)

Consider D. We know that they drew a match. This must have been v. E. And since D has 3 goals against, and the score in D's other match v. B was 4–0, ∴ D v. E was 3–3.

Complete Solution

A v. B : 0–1
A v. E : 3–2
B v. D: 0–4
B v. E : 1–0
C v. E : 4–0
D v. E : 3–3

64. 'What a way to end this song!'

We are given:

	PLAYED	WON	LOST	DRAWN	GOALS FOR	GOALS AGAINST	POINTS
A	2		0	2		7	2
B	3		0	I	5		5
C			2		2	2	2
D				2	7		2
E	4		3	I	4	I3	I

We must first look for the incorrect figure.
Consider E. We have:

PLAYED	LOST	DRAWN	POINTS
4	3	I	I

If *one* of these figures were changed – e.g. played 4 – then at least one of the other figures would have to be changed too.
Consider A. We have:

PLAYED	LOST	DRAWN	POINTS
2	0	2	2

Again if *one* of these figures were changed, then at least one of the other figures would have to be changed too. But we know that only one figure is wrong, . ˙. figures above for E and A must be correct.

E played all the other four, and both of A's 2 matches were drawn. . ˙. E v. A was a draw . ˙. E lost their other matches v. B, C and D.

Consider D. We are told that they drew 2 and got 2 points. But we know that D beat E. . ˙. either D's draws or D's points must be incorrect. . ˙. all other figures are correct.

Consider C. They lost 2 and had 2 goals against, . ˙. the scores in the 2 matches they lost were 0–1. And since A drew both their matches, and C beat E, . ˙. C v. B was 0–1, and C v. D was 0–1 . . ˙. C v. E was 2–0. And C did not play A.

. ˙. We know that D beat C and E . . ˙. D's points (2) must be wrong.

∴ everything else is correct. ∴ D drew 2, ∴ D v. A and D v. B were both drawn. ∴ D's points should be 6 (not 2).

And since A only played 2, A did not play B.

We know now who played whom and the results of each match. Thus:

	A	B	C	D	E	GOALS FOR	GOALS AGAINST
A	╳	╳	╳	Dr	Dr		7
B	╳	╳	W 1–0	Dr	W	5	
C	╳	L 0–1	╳	L 0–1	W 2–0	2	2
D	Dr	Dr	W 1–0	╳	W	7	
E	Dr	L	L 0–2	L	╳	4	13

Suppose E v. A were 3–3 (it cannot be more than 4–4). Then D v. A would be 4–4 (A had 7 goals against). *And suppose D v. B was 0–0.* Then D v. E would be 2–? (D had 7 goals for). And E v. D would be ?–2, and since E had 13 goals against, E v. B would be ?–6. But this is not possible, for B only scored 5 goals. And if E v. A were 3–3 and D v. B were 1–1, then E v. B would be ?–7. ∴ E v. A cannot be 3–3 or less, ∴ E v. A must be 4–4, and D v. A is 3–3.

Since E scored 4 goals v. A, they scored 0 v. B and D.

Suppose D v. B was 1–1. Then D v. E would be 2–0, and E v. B would be 0–5. But this is not possible, for B only scored 5 goals and they scored 1 v. C.

∴ D v. B can only be 0–0. ∴ B v. E was 4–0, and D v. E was 3–0.

Complete Solution

A v. D: 3–3
A v. E : 4–4
B v. C: 1–0
B v. D: 0–0
B v. E : 4–0
C v. D: 0–1
C v. E : 2–0
D v. E : 3–0

65. Two, Three, Four, Six

```
            —  4        (i)
  — 6 ) — — —           (ii)
        — —             (iii)
      ————————
        — 2 —           (iv)
        — — 3           (v)
      ————————
```

We must first find the incorrect figure: 6 in the divisor and 3 in (v) cannot both be right, for any multiple of an even number is even. And 4 in (i) and 3 in (v) cannot both be right, for any number multiplied by 4 must be even. But only one figure is wrong.

∴ it must be the 3 in (v).

∴ since 6 and 4 are right, the last figure in (v) should be 4.

∴ (iv) and (v) are —24, and this is the result of —6 × 4, (iv) and (v) must therefore be 2 2 4 and the divisor must be 56.

∴ (iii) is 56, and the first figure of (i) is 1.

Add up from the bottom and we get:

Complete Solution

```
       1 4
  5 6 ) 7 8 4
       5 6
       ————
       2 2 4
       2 2 4
       ————
```

66. Oh! Uncle, oh! Bungle

The facts given are as follows:

	PLAYED	WON	LOST	DRAWN	GOALS FOR	GOALS AGAINST	POINTS
A					1	1	3
B			2		4	5	0
C	2			2			
D					6	3	5

Look first for the incorrect figure.

If C played 2 and drew 2, then their goals for must be the same as their goals against. But other goals for (1 + 4 + 6) do not add up to the same as other goals against (1 + 5 + 3) (the total of goals for must always be the same as total of goals against, for each goal appears twice).

∴ *either* C's played (2), or drew (2), *or one* of the figures for goals for or against must be wrong. ∴ all other figures are correct.

Consider B. They lost 2, and since they got no points, ∴ they can only have played 2. ∴ their goals for should be at least 2 less than their goals against. But they are 4 for and 5 against. ∴ *either* B's goals for (4) *or* B's goals against (5) must be wrong. ∴ all other figures are correct.

Consider A. They got 3 points, and had 1 goal for and 1 against. ∴ they must have played 3 games, and scores were 1–0, 0–0, 0–1. ∴ A played all the others, and since C drew both their matches A v. C was 0–0. D got 5 points, ∴ they won 2 and drew 1. ∴ A v. D was 0–1. ∴ A v. B was 1–0.

Consider C. They played 2 and drew 2. But not v. B who got no points. ∴ C did not play B. And C v. D was a draw.

D got 6 goals for, 1 v. A, ∴ 5 v. B and C, and B and C scored 3 v. D.

Suppose D v. C was 0–0. Then D v. B would be 5–3; and B's goals for and against would be 3–6 (remember that B v. A was 0–1). But in this case B's goals for and against would both be wrong. ∴ D v. C was *not* 0–0.

Suppose D v. C was 2–2. Then D v. B would be 3–1; and B's goals for and against would be 1–4. But in this case B's goals for and against would both be wrong. ∴. D v. C was *not* 2–2. (And if D v. C was 3–3 or more the situation would be worse).

∴. D v. C can only be 1–1. ∴. D v. B was 4–2. And B's goals for and against were 2–5. ∴. the incorrect figure is B's goals for; it should be 2, not 4.

Complete Solution

The incorrect figure is B's goals for; it should be 2, not 4.

A v. B : 1–0
A v. C: 0–0
A v. D: 0–1
B v. D: 2–4
C v. D: 1–1

67. Logic Lane

The house has three different figures, and they 'rise', i.e. each one is greater than the one before. ∴. the number must be between 123 and 789, ∴. the sum of the digits is between 1 + 2 + 3 (i.e. 6) and 7 + 8 + 9 (i.e. 24). For the sum to be in the 'upper half of what you have in mind' it must be greater than 15.

We also know that the number has 2 different prime number factors both of which are greater than 13.

Leaving out even numbers and multiples of 5 the possibilities, with their factors, are:

169 (13 × 13)	369 (3 × 3 × 41)	489 (3 × 163)
179 (None)	379 (None)	567 (3 × 3 × 3 × 3 × 7)
189 (3 × 3 × 3 × 7)	389 (None)	569 (None)
259 (7 × 37)	457 (None)	579 (3 × 193)
269 (None)	459 (3 × 3 × 3 × 17)	589 (19 × 31)
279 (3 × 3 × 31)	467 (None)	679 (7 × 97)
289 (17 × 17)	469 (7 × 67)	689 (13 × 53)
367 (None)	479 (None)	789 (3 × 263)

The number of the house is therefore 589.

68. 'Do you add or subtract? That's for you to discover'

E	R	I	C		(i)
I	P	I	C		(ii)
R	V	E	E		(iii)

Suppose that we subtract. Then from the last column E would be 0. ∴ in the first column E would be 0, but subtraction would not then be possible for (ii) would be greater than (i). ∴ it must be an addition sum.

Consider the last column and the third column. From the last column E is even, ∴ there cannot be anything to carry to the third column. ∴ C = 1, 2, 3 or 4.

But since I and C are not the same, ∴ there must be something to carry from the third column. ∴ I = 6, 7, 8 or 9.

Consider the first column. E is at least 2, and I is at least 6. And if E were 4, then I would be 7, which is not possible. ∴ E = 2, C = 1, I = 6.

R in (iii) must be 8 or 9. *Suppose 9.* Then in the second column 9 + 1 (carried from the third column) + P = V. But this is not possible, for P and V would then be the same. ∴ R = 8, P = 0, V = 9.

Complete Solution

```
2 8 6 1
6 0 6 1 +
8 9 2 2
```

69. Sons and Grandsons

The position for grandfathers and their sons is as follows:

(i) $\left.\begin{array}{l} B \\ C \\ D \end{array}\right\} \rightarrow P, Q, T$

(ii)
$$\left.\begin{array}{l} A \\ B \\ C \\ E \end{array}\right\} \rightarrow P, R, S, U$$

(iii)
$$\left.\begin{array}{l} A \\ B \\ D \end{array}\right\} \rightarrow Q, R, S, T$$

(iv)
$$\left.\begin{array}{l} B \\ D \\ E \end{array}\right\} \rightarrow Q, T, U$$

In order to find the father of, say, P, we look first at the places where P is absent. And we see that P is absent in (iii) and (iv). ∴ we know that P's father is *not* A, B, D or E; ∴ it must be C.

Similarly Q's father is *not* A, B, C or E; ∴ it must be D.

And R's father is *not* B, C, D or E; ∴ it must be A.
And S's father is *not* B, C, D or E; ∴ it must be A.
And T's father is *not* A, B, C or E; ∴ it must be D.
And U's father is *not* A, B, C or D; ∴ it must be E.

And we see that B has no sons.

The position for P, Q, R, etc., and their sons is as follows:

(i)
$$\left.\begin{array}{l} Q \\ R \\ T \end{array}\right\} \rightarrow I, N, O$$

(ii)
$$\left.\begin{array}{l} P \\ R \\ S \\ U \end{array}\right\} \rightarrow J, K, L, M$$

(iii)
$$\left.\begin{array}{l} Q \\ T \\ U \end{array}\right\} \rightarrow I, K, N, O$$

(iv)
$$\left.\begin{array}{l} P \\ S \\ T \end{array}\right\} \rightarrow J, L, M, O$$

By using a similar argument we get:

I's father is *not* P, R, S, T or U;	∴ it is Q.
J's father is *not* Q, R, T or U;	∴ it is P or S.
K's father is *not* P, Q, R, S or T;	∴ it is U.
L's father is *not* Q, R, T or U;	∴ it is P or S.
M's father is *not* Q, R, T or U;	∴ it is P or S.

N's father is *not* P, R, S, T or U;   ∴ it is Q.

O's father is *not* P, R, S or U;   ∴ it is Q or T.

But we know that no one had more than two sons, and as Q has I and N as sons, he cannot also have O. ∴ O's father is T.

And as J, L and M have as father either P or S, ∴ two of them must have one and the third the other. ∴ P and S both have one or more sons.

∴ it is R who has no sons.

Complete Solution

Bert has no sons; Roland has no sons. Kenneth's grandfather is Ernie. Orlando's grandfather is Duggie.

70. Add Twice

(i)	h	m	p	d	m		(i)	—	—	—	—	—
	b	h	p	h	m			—	—	—	—	—
	r	c	d	h	a			—	—	—	—	—

(ii)	r	b	a	d		(ii)	—	—	—	—
	p	q	h	d			—	—	—	—
	a	a	d	d			—	—	—	—

It is important to have a pattern, as above, and to fill in the blanks as they are found.

Consider the fourth column of (ii) $\begin{array}{c} d \\ \underline{d} \\ d \end{array}$ *d can only be* 0.

From the first column of (i), $h + b + ? \, 1 = r$, ∴ r is at least 3. From the first column of (ii), $r + p + ? \, 1 = a$, ∴ a is at least $(3 + 4 + ? \, 1)$, i.e. $7 + ? \, 1$ (remember that all letters must stand for different digits). But from the fifth column of (i), a must be even, ∴ $a = 8$.

∴ from the third column of (ii) $h = 2$. And from the third column of (i) $p + p = 10$, ∴ $p = 5$. And from the fifth column of (i), $m = 4$, and from the second column, $c = 7$. From the first column of (ii), $r = 3$ or 2; but not 2, for $h = 2$. ∴ $r = 3$. ∴ from the first column of (i), $b = 1$. And from the second column of (ii), $q = 6$.

Complete Solution

(i) 2 4 5 0 4
     1 2 5 2 4
     ───────
     3 7 0 2 8

(ii) 3 1 8 0
      5 6 2 0
      ───────
      8 8 0 0

71. The Professor and the Islands of the World

The total of goals for must be equal to the total of goals against, ∴ B had 33 goals against.

C drew 1, but C's goals for are not equal to C's goals against, ∴ C played 2. And since neither A's nor B's goals for are equal to their goals against, which ever of them drew with C played 2 matches.

∴ they all played 2, for matches played must be even.

B had 26 goals for, and A and C had 22 and 23 goals against, ∴ the score in A v. C was 22 + 23 − 26, i.e. 19. But this cannot be a draw, for the total of goals should be even.

∴ C v. B must have been the match that was drawn.

B and C had 33 + 23 goals against, and A had 28 goals for. ∴ the goals scored in B v. C was 33 + 23 − 28, i.e. 28.

∴ B v. C was 14–14. ∴ C v. A was 10–9 (24–14 v. 23–14), and A v. B was 19–12 (28–9 v. 22–10).

Complete Solution

A v. B: 19–12
A v. C: 9–10
B v. C: 14–14